TERTIARY LEVEL BIOLOGY

Principles and Techniques of Contemporary Taxonomy

DONALD L.J. QUICKE
Department of Biology
Imperial College at Silwood Park
Ascot, UK

BLACKIE ACADEMIC & PROFESSIONAL
An Imprint of Chapman & Hall

London · Weinheim · New York · Tokyo · Melbourne · Madras

Published by
Blackie Academic & Professional, an imprint of Chapman & Hall,

Chapman & Hall, 2—6 Boundary Row, London SE1 8HN, UK

Chapman & Hall Gmbh, Pappelallee 3, 69469 Weinheim, Germany

Chapman & Hall USA, 115 Fifth Avenue, Fourth Floor, New York, NY 10003, USA

Chapman & Hall Japan, ITP-Japan, Kyowa Building, 3F, 2-2-1 Hirakawacho, Chiyoda-ku, Tokyo 102, Japan

DA Book (Aust.) Pty Ltd, 648 Whitehorse Road, Mitcham 3132, Victoria, Australia

Chapman & Hall India, R. Seshadri, 32 Second Main Road, CIT East, Madras 600 035, India

First edition 1993
Reprinted 1996 (twice)

© 1993 Chapman & Hall

Typeset in 10/12 pt Times New Roman by
Columns Design & Production Services Ltd, Reading
Printed in Great Britain by St Edmundsbury Press Ltd, St Edmunds, Suffolk

ISBN 0 7514 0019 X (HB)
       0 7514 0020 3 (PB)

A catalogue record for this book is available from the British Library

Library of Congress Cataloging-in-Publication Data Available

∞   Printed on acid-free text paper, manufactured in accordance with
      ANSI/NISO Z39.48-1992 (Permanence of Paper)

Preface

Taxonomy is an ever-changing, controversial and exciting field of biology. It has not remained motionless since the days of its founding fathers in the last century, but, just as with other fields of endeavour, it continues to advance in leaps and bounds, both in procedure and in philosophy. These changes are not only of interest to other taxonomists, but have far reaching implications for much of the rest of biology, and they have the potential to reshape a great deal of current biological thought, because taxonomy underpins much of biological methodology. It is not only important that an ethologist, physiologist, biochemist or ecologist can obtain information about the identities of the species which they are investigating; biology is also uniquely dependent on the comparative method and on the need to generalize. Both of these necessitate knowledge of the evolutionary relationships between organisms, and it is the science of taxonomy that can develop testable phylogenetic hypotheses and ultimately provide the best estimates of evolutionary history and relationships.

Because of its breadth, taxonomy is a complex mixture of biology, philosophy and mathematics. It should therefore come as no surprise to find that to understand the whole of taxonomy would require an extensive knowledge of both the techniques of data acquisition and a bewildering array of analytical procedures. However, in many parts of the world, rather than explaining these aspects to undergraduates, university biology courses have been progressively marginalizing taxonomy in favour of more 'trendy' areas. This book aims to stem this trend, by combining up-to-date arguments about theory with a survey of the kinds of data that taxonomists can put to use.

The first part of this book introduces the student to the range of considerations about characters, taxa and techniques that forms the basis of good taxonomic practice and, in particular, the interpretation of likely evolutionary relationships. A brief summary of the current rules of nomenclature as they pertain to plants, animals, bacteria, etc., is provided, although this should be considered only as a summary. The second part of the book covers in more detail the range of techniques used to collect taxonomic characters, whether cytological or molecular. The last chapter assesses the relevance of taxonomy to studies of biodiversity and conservation, and surveys the roles of taxonomic institutions, such as the major museums, in this work.

With the current rapid advances in DNA sequencing and even the extraction and sequencing of DNA from fossils, the next few years seem likely to be marked by many remarkable discoveries in taxonomy no less so than in other areas of science.

Acknowledgements

I would like to thank Trevor Elkington, Mike Fitton and Steve Rolfe for many helpful discussions during the preparation of this book, David Hollingsworth for his assistance in preparing the cover illustration and Rachel Kruft, who carefully proof-read much of the original manuscript. Above all, I wish to express my thanks to my wife, Tanya, and my parents, Walter and Maisie, for their great support and tolerance of my work over many years.

D.L.J.Q.

Contents

Chapter 6   NOMENCLATURE AND CLASSIFICATION   115

Chapter 7   CYTOTAXONOMY   136

CHAPTER ONE
INTRODUCTION

The field of systematics has been in considerable turmoil as various investigators developed different methods of classification and argued their merits. I guarantee you that no one method or view has all the good points.

Fitch (1984)

1.1 The compass of taxonomy and systematics

There is no one definition either of **taxonomy** or of **systematics**. Indeed many workers have used these two terms more or less synonymously. Yet, it is possible to make a valid and useful distinction between them as, for example, is suggested by Hawksworth and Bisby (1988). Taxonomy in this sense includes a range of different areas from the description and naming of new taxa (**nomenclature**), the arrangement of organisms into a convenient classificatory system (**classification**), and the construction of identification systems for particular groups of organisms. Systematics, which is really the subject of this book, may be considered as a rather broader topic of which taxonomy is only a part, albeit a substantial one. Thus systematics includes traditional taxonomy with the addition of theoretical and practical aspects of evolution, genetics and speciation. It is also often helpful to identify separately the explicit study of the evolutionary relationships between organisms. This aspect of systematics is usually referred to as **phylogenetics**.

Taxonomy and systematics encroach either directly or indirectly on many other areas of science including fields as diverse as agriculture, horticulture, medicine, pharmacology, anthropology, archaeology and petrology as well as the traditional areas of botany, zoology and microbiology. Within each of these, they provide names for organisms and a framework within which these are classified. On top of this, the hypotheses that systematics generates about the evolutionary relationships between organisms form the basis of the comparative method, a

central technique for drawing unbiased conclusions about the relationships among characters and between characters and environmental
factors (see section 1.5).

Unfortunately, despite their obvious importance, taxonomists have
long drawn a measure of scorn from other biologists who widely
perceive taxonomic work as a rather imprecise area of research.
Indeed there is the well-known tautological definition that a species is
that which a *good* taxonomist says is a species, and so on. Implicit in
this adage is the notion that when an experienced taxonomist is
presented with conflicting sets of data, he or she will usually be able to
select those characters that will give an evolutionarily correct result.
This methodology in fact accounts for the great majority of taxonomic
descriptions and decisions that have been made to date, although of
course not all practising taxonomists live up to the almost divinely
guided image invoked.

The truth is that the role of the taxonomist has all too often been
misunderstood by other biologists, and few who are not directly
involved have any real concept of the rigour involved in the proper
execution of taxonomy. Without doubt the situation has not been
helped by the current and widespread downturn in taxonomic
research, and more worryingly in training, in favour of a number of
more trendy areas of enterprise (see chapter 13). Ironically, this
downturn has occurred at a time when a combination of circumstances
have made the need for more taxonomic research urgent. Firstly, the
destruction of many natural habitats giving rise to the so-called
biodiversity crisis means that many organisms are in danger of going
extinct before they can be described and classified. Secondly, rapid
advances in both taxonomic methodology and molecular biology are
making possible the rapid acquisition of potentially enormous amounts
of new data that can be used in trying to unravel the evolutionary
histories or **phylogenies** of extant organisms.

1.2   The 1960s and the emergence of new ideas

Many researchers have argued that the traditional subjectivity of
taxonomy does not form the basis of true science and that clearly
defined criteria ought therefore to be applied to taxonomic decisions
just as they appear to be in other areas of research. More objective
criteria might be used, for example, in determining whether or not a

specimen belongs to a new taxon, what relationships exist between the members of a group of taxa, or where to draw dividing lines between higher taxa.

In the 1960s, the call for a more objective taxonomy led to the largely separate development of two new and more rigorous areas of systematics. One of these, known from its inception as **numerical taxonomy** but nowadays frequently and perhaps more appropriately referred to as **phenetics**, was largely developed and popularized by Sneath and Sokal (1973). As the name implies, numerical taxonomy consists of applying various mathematical procedures to numerically encoded character state data for the organisms under study. The products of these operations were often taken to be 'unbiased' indicators of the similarity or difference between the taxa, which were in turn used to arrange taxa in a hierarchy.

For some years, the application of phenetic versions of numerical taxonomy in systematics research became virtually *de rigeur* as have many other fashions in science. However, the apparent rigour provided by these early mathematical procedures did not necessarily yield answers that were particularly better than those achieved previously by the traditional subjective approach. Further, it is probably unreasonable to expect the majority of people to interpret results that were typically presented as dendrograms as anything other than suggestions of evolutionary histories even if, as was the case, most proponents of numerical taxonomy went out of their way to point out that their trees were not intended to depict phylogeny. Unfortunately though, as many people frequently but unjustifiably did interpret the results as indicators of phylogeny, the results could even be said to be misleading.

At about the same time as the development of numerical phenetic methods, a series of very different works were being published by the German systematic entomologist Willi Hennig, and these were ultimately to lead to a revolution in taxonomic thinking (Hennig, 1950, 1957), although their impact was only fully realized when they became more widely known outside of Germany largely through translation into English (Hennig, 1965, 1966, 1975), and in particular their initial spread was most marked in Hennig's own field, entomology (Dupuis, 1984). Hennig's vital contribution to modern systematics was the combination of emphasizing that phylogenetic reconstruction can only be based on shared **derived** features and his provision of a methodology for phylogenetic analysis based on this principle.

Hennig's methods and their subsequent derivations have come to be known as **cladistics** or **evolutionary systematics** because of their specific claim to be able to reveal the evolutionary relationships between the members of a group of taxa.

The Hennigian method was initially a largely logical exercise and, as personal computers were only just on the horizon, analyses using this method were at first largely restricted to a small number of taxa and characters. Surprisingly then, whilst the logic of Hennig's method was ideally suited to computerization in order to analyse large data sets, early applications were largely done by hand. Indeed as pointed out by Felsenstein (1988), a split soon developed between followers of Hennig's **phylogenetic systematics** and workers involved in computer-based numerical analyses even though the latter camp included both **pheneticists** and others such as Edwards and Cavalli-Sforza (1964) and Camin and Sokal* (1965) who were in fact developing algorithms based on **parsimony** that would ultimately prove vital to the furtherance of cladistics.

1.3   Cladistics and numerical taxonomy: the conflict

The phenetic approach of numerical taxonomy is based on the assumption that the more similar two taxa are, then the more closely related they are likely to be. Although at first sight this might appear to be a reasonable assumption, on closer inspection this can be shown to be invalid.

In numerical taxonomy, character states are each assigned a numerical value and the resulting sets of values (data matrices) obtained for several characters over a group of individuals or taxa is then processed with the aim of using them to obtain unbiased taxonomic inferences. For example, by use of an appropriate algorithm it is possible to use the numerical data to define groups of **OTUs** (operational taxonomic unit) based on overall similarity, a process referred to as **phenetic clustering** (see section 4.3). Sokal and Sneath (1963) provide an extensive discussion of their reasons for believing that phenetic clustering techniques are superior to cladistic methods

---

* Camin and Sokal (1965) although describing an essentially cladistic technique were not particular supporters of cladistics and in their paper (Camin and Sokal, 1965) they suggest that phenetic methods will probably prove preferable for use as a basis for classification.

for forming the basis of a classification. Briefly, they argue that natural (i.e. **monophyletic**) groups may not share any one single defining character as it is often purely a matter of chance whether or not a member of the group will show a character reversal such that it loses the diagnostic feature. For this reason, Sokal and Sneath advocate the use of a **polythetic** system for defining classificatory groups such that possession of at least some minimum number of a set of characters will justify placement of a taxon in that group while not requiring that any included taxon must display all the character states in the polythetic set. Through this reasoning they therefore propose that higher taxa can be defined through the use of cluster analysis such that included members will resemble one another more on average than they will non-members.

The realization that numerical taxonomy does not provide a satisfactory way of identifying relationships started with the work of the German systematist Willi Hennig who made the apparently simple assertion that classifications should reflect the evolutionary history of the group and then critically considered what actually happens to character states during evolution. Hennig noted that during evolution, characters evolve from one state (the primitive or **plesiomorphous** state) to another (the advanced, derived or **apomorphous** state). Therefore, he argued, within a group of related organisms in which a character shows more than one state, only one of these can be the ancestral state and if this is so then all of the others are derived (apomorphous) states which have evolved within the group from that ancestral state. The significance of this to reconstructing evolutionary trees and hence to creating natural classifications can be seen by considering the hypothetical dataset shown in Figure 1.1. Here the states of five binary characters (coded as 0s or 1s to represent plesiomorphous and apomorphous states, respectively) are shown for each of three taxa, A, B and C. As shown in the lower part of Figure 1.1, in this and many other examples application of the numerical taxonomic technique of phenetic clustering leads to a different tree than Hennig's cladistic method, but only one of these can be a true representation of their evolution. The reason for this conflict is easy to see. The phenetic tree reflects the fact that taxa A and B share more character states in common than either does with taxon C, but in each case the shared feature is plesiomorphous, that is it would have been present in the group's common ancestor. In contrast, the cladistically constructed tree shows taxa B and C to be most closely related even though it is based on just one apomorphous character state that is shared by B and C; no other pairs of taxa share any derived features.

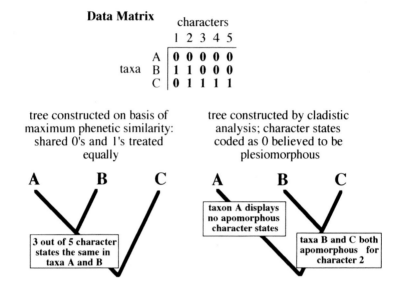

Figure 1.1 A simple data matrix with five characters and three taxa is used to demonstrate how a tree constructed on the basis of overall phenetic similarity (a phenogram) (*left*) can differ from one based on cladistic principles (cladogram) (*right*) which makes use of the polarity of character states. (Modified after Sytsma, 1990.)

From the foregoing demonstration, it would be easy to assume that the death knells of numerical taxonomy as a whole started to sound with the publication of Hennig's work. However, this is not entirely so and it would be wrong to lump all numerical techniques under the same umbrella as phenetic clustering. Thus while many of the taxonomic techniques developed by workers such as Sokal and Sneath are now largely defunct or redundant, some still have a great deal of value for taxonomic research and these will be discussed in chapter 4.

1.4   Assumptions and philosophy of cladistics and the use of parsimony criteria

In considering the development of cladistic methodology in taxonomy, not only is the continuing development of new techniques notable but also the continuing, and at times acrimonious, debate between proponents of different schools (Felsenstein, 1988). Papers involving particular techniques may even be effectively barred from some journals whose editorial board hold conflicting views.

First of all, it should be pointed out that there is no firm consensus about what exactly constitutes the Hennigian methodology that underlies modern cladistics. Readers interested in this historical debate are referred to a series of papers which attempt to examine what exactly Hennig was doing in his various examples (Felsenstein, 1983; Duncan, 1984; Farris, 1986). What is important is that like most other scientific methodologies, phylogenetic systematics has evolved considerably since it was first conceived (Platnick, 1979) and, equally importantly, it is still developing at a pace evidenced by the large numbers of papers published on cladistics every year.

The development of computing in taxonomy has been intimately linked with the development of cladistic methodology over the past 25 years or so. One major consequence of the computerization of Hennigian cladistics was that there was a need to specify exactly how a Hennig-based algorithm should function and specifically to define exactly what answer was being sought. A second consequence was that workers wanted to make use of the power of computers to cope with ever larger data sets, and this inevitably led to the inclusion of characters that might be rather less than 100% reliable indicators of phylogeny. In other words, characters that must have either been mis-scored as **homologous** (due to parallel evolution or convergence; see chapter 2) or have undergone reversals back to their ancestral state. These incongruences, collectively referred to as **homoplasy**, lead to conflicting evolutionary hypotheses.

In Hennig's original method, relatively few apparently incongruent characters would be employed and if two characters were found to disagree with one another, Hennig's solution was to study each character until the reason for the disagreement could be identified and corrected (Meacham and Estabrook, 1985). Computers, however, are not yet able to re-examine characters and so have to accept the input given to them by their operators. Essentially two possible solutions were initially proposed. One is based on the idea that all **true characters**, that is ones which truly reflect phylogeny, must be consistent with one another. The other, which has been far more widely accepted, is based on the idea that the best hypothesis about relationships based on a given data set is the one which necessitates the fewest evolutionary changes. Techniques based on the first of these ideas are referred to as **compatibility** techniques (see section 3.1.2) while those based on the second are termed **parsimony** techniques (see section 3.1.1).

Sober's (1988) critical appraisal of many of the logical and philosophical issues involved in cladistic analysis, and in particular the assumption that the simplest (most parsimonious) hypothesis is necessarily the best hypothesis, comes to some disconcerting conclusions. Many of the arguments that have been postulated in favour of parsimony as the best criterion for choosing between alternative phylogenetic hypotheses are shown to be flawed. Equally though, many arguments postulated against parsimony are also flawed. So what does this mean? Probably primarily it points to the fact that few biologists who are the users and appliers of cladistics have any, let alone a thorough, training in philosophy. Secondly, it may mean that there is no one answer to the question. In some circumstances parsimony may be the best criterion, in others it might not. This begs the question as to what exactly it means for evolution if, in fact, parsimony is the best method for inferring phylogenies? Does it, for example, mean that evolution is itself a parsimonious process? The answer to this last question is probably not, but it is not clear whether this invalidates parsimony as a criterion for estimating phylogeny.

Luckily, as pointed out by Sober, the failure of arguments used in favour of parsimony does not mean that parsimony itself is not the best criterion for phylogenetic analysis.

1.5 Taxonomy and the comparative method in biology

It is becoming increasingly realised — perhaps for the second time around — that taxonomy, and phylogenetic studies in particular, can add greatly to other areas of biology. Well-founded phylogenetic hypotheses can be used to generate experimentally testable predictions about the evolution of many biological systems from biochemical and physiological to behavioural studies (McLennan *et al.*, 1988). However, the key point in this argument is that the phylogeny should be as accurate as possible since inaccuracies could lead to totally wrong conclusions. Thus one of the most important contributions that phylogenetic studies can make to other branches of biology is that they permit the separation of potentially confounding phylogenetic factors from studies of comparative biology. As such phylogenetic interpretation is crucial to the so-called **comparative method**, first introduced by Darwin. By use of the comparative method the true significance of correlations, for example betwen biology and ecology, can be discerned (Felsenstein, 1985a; Wanntorp *et al.*, 1990; Brooks and McLennan, 1991; Ross and Carpenter, 1991; Harvey and Pagel, 1991).

To illustrate the power of the comparative method, consider a typical sort of biological observation such as a correlation between the number of larval instars shown by parasitic wasps and whether or not they are endoparasitoids or ectoparasitoids. This correlation with endoparasitoids going through fewer larval stages could well be indicative of some important biological processes perhaps concerned with strategies for evading host defence mechanisms, and as such could well justify further investigation. Alternatively, however, the correlation could be due purely to chance and apparently magnified in significance because all the wasps displaying the smaller number of larval stages happen to comprise a single evolutionary lineage that just happen to be endoparasitoids. Which of these alternatives one prefers will be influenced by how many separate occasions that endoparasitic wasps have evolved to have fewer instars, but in order to know this, it is necessary to know the phylogeny of the group of wasps concerned. The more distinct lineages of endoparasitic wasps that have independently evolved to have fewer instars, the more likely it is that some causal mechanism exists between the two biological characters.

This argument can also be employed in reverse at the experimental design stage. Thus, for example, to discover whether differences in some biological trait reflect a particular selective pressure, then either searches can be made for large sale correlations involving multiple instances of parallel or convergent evolution (as above) or alternatively, comparisons could be made between members of fewer but carefully selected pairs of closely related species that are therefore likely to be genetically similar but differ in the critical biological feature under scrutiny. The discovery of consistent pairwise correlations under such conditions would provide perhaps the strongest evidence in favour of some causally determined connection. Through the careful phylogenetically determined selection of study species, meaningful results may be obtained at far less expense of time and effort than they would be by working on randomly selected taxa.

In addition to helping to avoid phylogenetic bias in data interpretation, phylogenetic hypotheses may also be used to help unravel other aspects of biological evolution. These roles include working out the evolutionary sequence involved in the development of a trait in the evolution of a complex behaviour pattern, or possible limitations and necessary preconditions for the evolution of a new feature. For example, Eggleton (1991) in an investigation of the relationship between mating strategy, ecology and phylogeny in a group of parasitic wasps has provided evidence that ecological pressures are not sufficient to explain behaviour and that there may also be direct phylogenetic limitations. Similarly, Albert *et al.* (1992) have recently used a phylogenetic approach to show that both carnivory and particular types of insect-trapping mechanism have evolved on several independent occasions among the plants. Further, phylogenetic reconstructions can also be used to generate predictions about the evolution of behaviours which may then be tested experimentally (McLennan, 1991). The ability of independently constructed phylogenies to generate hypotheses about the evolutionary changes that have taken place in other character systems including ecological and ethological features, is likely to be one of the great strengths of cladistic analysis in the future.

CHAPTER TWO

CHARACTERS, TAXA AND SPECIES

Biology has long endured arguments over species concepts. The reasons
for this contentiousness are not simple, but the willingness of biologists
to engage in continued intellectual debate underscores the critical
importance placed on species and the fact that no solution to the 'species
problem' has been generally accepted.

Cracraft (1989)

2.1 Nature and handling of data

This chapter covers some of the many considerations that have to be
taken into account when assembling data for taxonomic usage.
Considerations of what characters are and how they should be encoded
for analysis, and what taxa are and how to distinguish between natural
and artificial groupings are vital aspects of any phylogenetic analysis as
well as having major implications for phenetics, classification,
nomenclature and many other areas of biology and taxonomy.

In terms of characters, particular attention has to be given to the
ways different types of character are encoded, how primitive and
advanced character states may be recognized, how to distinguish
similarities resulting from common ancestry from the effects of
convergent or parallel evolution, and in what ways one character state
is likely to evolve into another. In any cladistic analysis such
considerations have to be applied to each and every character.

2.2 Characters

The raw data of both numerical taxonomy and phylogenetic methods
are generally taxa versus character matrices which can be manipulated
in many different ways according to requirements. In this case the
word **character** is used to refer to a particular attribute of a specimen

that can take two or more different forms or **character states** (Platnick, 1979). Thus to quote Watrous and Wheeler (1981) '. . . a character is an original form plus all of its subsequent modifications'.

Before any data analysis can be started, some consideration needs to be given to how exactly character states are encoded, particularly as not all characters are equally amenable to this procedure. Nearly all cladistic methods rely on discrete characters, that is ones which either intrinsically comprise a number of discrete states, such as the number of anthers in a flower, or segments in an insect's leg, or ones which can be coded for analysis purposes into a system of discrete states. Each of the possible states exhibited by a given character is then typically coded as a particular integer number, 0, 1, 2, etc. For example, if members of a group of plants display either white or blue petals, then white petals could be scored 0, and blue ones as 1, or vice versa.

The simplest and easiest characters to use and interpret are binary ones, that is ones with only two discrete states, such as the presence or absence of wings or such like. Often though, characters are far from easy to interpret. They may, for instance, be continuously variable or have more than two character states which means that consideration has to be given to what changes between states are permissible. A genus of plants, for example, may have species with red, white and blue flowers. In this case, red petals might be assigned the code 2, but this could be taken to imply that red petals are more different from white petals (assigned code 0) than are blue ones (assigned as code 1). In other words, if the numerical codes in this example were treated arithmetically, that coding system would imply a distinct evolutionary model which may or may not be desirable or intended. Such a model is usually referred to as **additive** and the characters as **Wagner** characters (see Figure 2.5).

In addition to simply assigning numerical codes to each character state, therefore, anyone wishing to analyse character state distributions from an evolutionary perspective must pay careful attention to character state transitions. A further aspect of character analysis essential to the application of Hennigian cladistics is the polarization of characters, i.e. the need to be able to distinguish ancestral (**plesiomorphous**) from derived (**apomorphous**) character states. These aspects of character analysis will be discussed in more detail in the following sections.

Even with characters that seem to comprise a number of discrete states, things may not always be as easy as they seem. Pogue and

Mickevitch (1990) have postulated that for simplicity many workers have tended to massage or even force observed variation in a character system into too small a number of discrete character states. They claim that by employing finer distinctions between character states the result is more realistic, although it necessitates a greater emphasis on character state transitions which in turn can be a major source of ambiguity in the analysis. They also propose a possible way around this problem which they term transformation series analysis.

2.2.1 Discrete coding of continuous characters and ratios

Continuously variable characters may be employed for cladistic analysis but they then need to be subdivided into two or more non-overlapping ranges (discrete character states). Unfortunately, very little is known about what is the best way to convert continuously variable characters into discrete ones despite the frequency with which taxonomists need to do this. Consequently it is a source of considerable disagreement and there may in fact be no one best way of coding continuously variable data. To determine the optimal strategies for particular types of data would probably require considerable simulation modelling in conjunction with the analysis of real data sets so that the effects of different evolutionary and population genetics models could be taken into account. However, funding for such research is hard to obtain.

In practice, many taxonomists are not particularly rigorous when it comes to the coding of continuous data, and arbitrarily chosen cut-off points between ranges are often employed. Of course, if the raw data do fall into two or more widely different and discrete ranges of values there may be nothing wrong with this process, but data are seldom so obliging. In these cases a more rigorous system of coding is called for and a number of more or less objective criteria have been proposed at various times, each probably having its own advantages and disadvantages. There have in fact been rather few investigations of the merits or otherwise of different approaches and it is likely that no one system will be optimal for all purposes. Several techniques were analysed by Thorpe (1984) including **gap-coding**, **segment coding**, and **divergence coding**. In the first of these some statistical measure of the variance of the data is used as a yard-stick for identifying *sufficiently* large gaps in the raw data. In segment coding, the total range of variation in a character is simply divided into an arbitrarily chosen, number of equal-sized, non-overlapping ranges. Divergence coding, differs from both gap and segment coding in that it makes use of prior knowledge of the taxa* from which the data are derived. The mean character values for each taxon are arranged in rank order and a taxon by taxon matrix is filled with either 1s, −1s or 0s depending on whether the means for each pair of taxa are significantly different (greater or less than) or not. Having completed this process, the total score of 1s, −1s and 0s is calculated for each taxon and that value used as the character state. Because of its reliance on statistical significance, an essential requirement of divergence coding is that each (or at least a large proportion of) taxa are represented by a reasonable sample size. Archie (1985) introduced a **generalized gap coding** procedure which should not be confused with gap coding *sensu* Thorpe; Archie's method, which is more akin to divergence coding than to gap coding, has been discussed by Farris (1990). In generalized gap coding, taxa are again arranged in order according to their mean values and *pooled* data for all possible sets of contiguous taxa are compared statistically with data for adjacent taxa or contiguous sets of adjacent taxa. If a statistically significant difference is obtained then the boundary between the two groups is taken as a 'gap'.

Baum (1988) has specifically addressed the problem of applying discrete coding that is applicable to cladistic analysis to continuously variable characters. His procedure involves ranking taxa according to the minimum value of the character they display and if ties occur then ranking tied taxa according to the mean or the mid-points of their ranges. The rank orders are then used as the character codings. He further suggests that in the case of data collected from museum specimens it might be wise to exclude the upper and lower extremes of the character ranges (e.g. omitting the ten percentiles) because of the frequent habit of collectors of collecting extreme examples! For analysis, Baum's codings require an additive interpretation (i.e. they should be treated as ordered characters). Also because the number of codes employed for different characters may be widely variable, character codes may need rearranging so that each character contributes equally to tree length. It should be noted that this is different from this author's view of how intrinsically discrete character states should be treated (see chapter 3, section 3.2.6).

Farris (1990) has provided a fairly detailed critique of generalized

---

* Recognition of what are taxa in the first place, however, is usually the result of more *ad hoc* processes of traditional (intuitive) taxonomy.

gap coding although his discussion is relevant to all methods. The problems that he highlights basically reflect the difficulty of ensuring that the coding procedure gives appropriate weight to particular degrees of overlap or distinctness in character ranges. For example, Baum's method gives an equal weight (different coding) to every different range even though many of the differences may simply be the result of sampling error. Divergence and generalized gap coding are likewise subject to problems due to sample size such that some significant variation will not show up as being significant in a statistical test. Further, because the latter two procedures assign importance to insignificant differences they tend to obscure truly important differences since these are given no more weight than any of the unimportant ones. Overall, the problem of coding continuous characters is far from simple, and it should always be realized that when continuous variation is encountered, it can result from many different underlying genetic mechanisms further confounded by environmental factors.

Another common problem in using measurements in taxonomy is the confounding effect of a real difference in size between the organisms under study, either within or between taxa. If the taxonomist is concerned with relative size of a particular feature then taking its absolute measurement is likely to be next to worthless. The usual way around this is to use the ratio of the size of the structure to some standard measurement, for example, wing vein lengths in an insect could be related to total wing length. However, this process is not always without its problems because a ratio may be large either because the character under consideration is also particularly large, or because the structure being used for comparative purposes is especially small. Thus, variation in both measurements need to be viewed in a wider context if the full significance of the data is to be realized. Another problem frequently encountered with the use of ratios is that the shape of a structure often changes through an organism's development and therefore many ratios that are intended to express something about a structure's shape will also be size-dependent. Phillips (1983) suggests one way of overcoming this by plotting the ratio against size and then using the resulting regression coefficient or intercept as the character rather than the value of the ratio itself. Of course, such procedures are only applicable in situations in which there are a sufficient number of specimens from which to calculate the regression.

2.2.2   *Identifying primitive and advanced character states*

As explained in the previous chapter, Hennigian cladistics and the various approaches to phylogeny reconstruction that it has given rise to require not only that characters are scored for their different states but also that for as many characters as possible a decision is made as to which of its states is the primitive condition that was displayed by the common ancestor of the group members. The term **plesiomorphous** is preferred to the term primitive because the latter carries with it additional connotations of simplicity, and because a character state that was displayed by the common ancestor of a group may be advanced compared with the state displayed by its own ancestors. Likewise, the term **apomorphous** is preferred to advanced. Both apomorphous character states and plesiomorphous ones can be shared by more than one species or other taxon, and two additional terms are employed to describe these situations. When a homologous apomorphous character state is shared by two or more groups (indicating that they may have a common ancestor) the character state is said to be a **synapomorphy** for the taxa possessing it. In contrast, if two or more taxa display the same plesiomorphous state for a character it is described as a **symplesiomorphy** for those taxa.

In the absence of a truly complete fossil record for any group, it is immensely unlikely that we can ever know ancestral features with absolute certainty. Therefore, we have to make **polarity** decisions based on the best interpretation of available evidence. This then poses the question, what criteria should be employed to make these decisions? Many criteria have been used at various times and in various circumstances to infer character polarities, and there have been a number of review articles dealing with the relative validities, merits or otherwise of these (Estabrook, 1977; Crisci and Stuessy, 1980; De Jong, 1980; Stevens, 1980; Arnold, 1981; Stuessy and Crisci, 1984). Although some suggested criteria have failed to stand the test of time (and incisive thought) there are still several in current usage and a few of the more commonly employed will be briefly reviewed below.

Probably the most reliable method of character state polarization, and nowadays also the most widely used and accepted of the various polarizing procedures, is the **outgroup method**. This has been thoroughly discussed by Watrous and Wheeler (1981) and only an outline is provided here. The underlying principle stems from the following argument. Imagine an ancestral group in which a taxon shows only one character state, and this taxon evolves to form two daughter groups. A mutation occurring in one of the daughter groups will lead to some of its included taxa showing a different character state from the remainder of that group and so that group will now appear polymorphic for the character concerned (Figure 2.1). Our problem is that we probably cannot know the character state of the ancestor itself, but we may be able to see the character state of the other daughter lineage. The latter, if it shows only one of the character states possessed by the group under consideration, will most likely show the primitive condition. Any other hypothesis is less than maximally parsimonious since it would require an additional, independent evolution of an indistinguishable apomorphous condition in the two daughter lineages. In this analysis then, the group whose

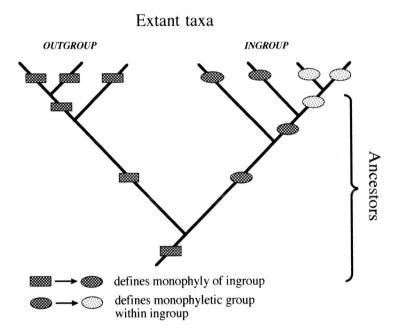

Figure 2.1 An illustration of the outgroup method for determining the plesiomorphous (ancestral) state of a character. The three outgroup taxa display the character state 'chequered' while two of the ingroup taxa display 'chequered' and two display 'stippled'. 'Chequered' rather than 'stippled' is considered to be primitive (plesiomorphous) as the alternative would require two character state changes from 'stippled' to 'chequered' rather than just one to explain the character state distributions among the extant taxa.

internal phylogeny is under investigation is termed the **ingroup** and its sister lineage, the **outgroup**.

More recent expansions of the outgroup analysis method increase its usefulness in cases when there may be several potential outgroups that, inconveniently, display more than one of the character states present in the ingroup (Donoghue and Cantino, 1984). The methodology developed by Maddison *et al.* (1984) makes use of independent (local) phylogenetic analysis of the outgroup taxa using parsimony (see chapter 3) in order to determine the most likely character states of their common ancestor. The greater the certainty about the relationships between the outgroup taxa, the greater will be the accuracy of the character polarization and hence of the final **cladogram** for the ingroup taxa. Maddison *et al.* also suggest the best approach to use when the phylogenetic relationships within the outgroup cannot be resolved with confidence.

In some circumstances, however, it may not be possible to apply the outgroup technique, perhaps because of the absence of homologous characters in the outgroup or an inability reliably to identify an outgroup. In these cases other criteria need to be employed. Among other sources of inference having various degrees of value can be listed commonality, co-occurrence of primitive states, extents of geographic distributions, fossil record, early developmental pattern (ontogeny), vestigial organs and *minor* developmental abnormalities.

Of these, the most widely used and probably most widely abused argument relates commonality to primitiveness, the so-called **commonality principle**. The principle behind this conjecture is illustrated in Figure 2.2, in which it can be seen that in the set of three possible unlabelled rooted trees for five terminal taxa, there are a total of seven possible evolutionary branches along which an apomorphous character state can evolve so as to yield phylogenetically useful information, i.e. each state is represented in at least two taxa. Out of these, the apomorphous state will be displayed by a minority of taxa in five instances, and by a majority in only two. Similar ratios are found for larger numbers of taxa and also if the different tree topologies are weighted in accordance with their probability of evolving (see chapter 12). The tendency towards 'common equals primitive' is even more marked if situations in which one character state is only displayed by a single taxon are taken into account. Thus, all other things being equal, commonality will *on average* equate to primitiveness (Frohlich, 1987).

The three possible tree topologies for 5 taxa

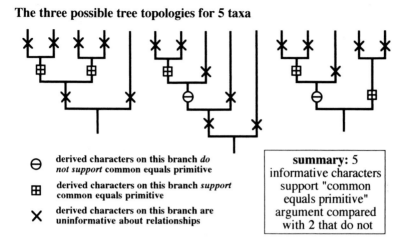

⊖ derived characters on this branch *do not support* common equals primitive	**summary:** 5 informative characters support "common equals primitive" argument compared with 2 that do not
⊞ derived characters on this branch *support* common equals primitive	
✗ derived characters on this branch are uninformative about relationships	

Figure 2.2 Demonstration of principle underlying 'common equals primitive' argument of character polarization using a five taxon example. The method works *on average* but is not deterministic and, as shown in Figure 2.3, tends to fail particularly badly the more comb-shaped the phylogeny is.

In an important paper, Watrous and Wheeler (1981) deal at some length with the flaws in this method, pointing out for example that it does not work if there are only three taxa, and showing that if the true phylogeny is rather comb-shaped then the commonality principle will lead to trees which are more symmetrical (Figure 2.3). However, the main issues are that commonality is probabilistic rather than deterministic and that its effectiveness is heavily dependent on the topology of the true evolutionary tree.

Partly related to the commonality principle are two other probabilistic concepts. First, there is the often stated idea that primitive character states will tend to co-occur in the same taxa (especially if the group is not too ancient). However, this presents the problem of how to distinguish between co-occurring primitive features and co-evolved advanced ones. Second, there is the idea that since older taxa will have had longer to disperse, then the most widely dispersed taxon is likely to display the most primitive character states. Exceptions to both of these are all too easy to imagine and it would be foolish to put much reliance on either of them.

The fossil record, were it complete, would almost certainly be able

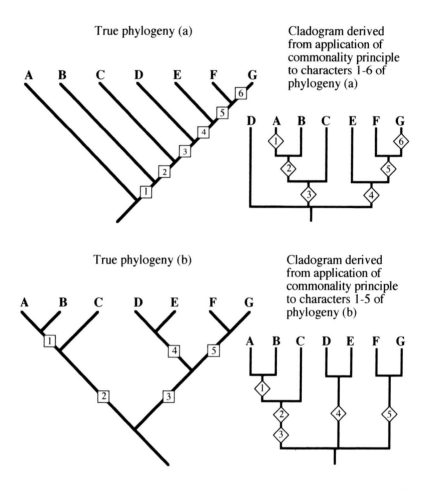

Figure 2.3 Diagrams illustrating the dependence of the success of the 'common equals primitive' method of character state polarization on the (unknown) phylogeny of the group being studied. The method is particularly unsuccessful if the true phylogeny is comb-shaped.

to reveal primitive character states because a complete record would enable construction of a near perfect phylogenetic tree from which primitive character states could then be inferred. Unfortunately, the fossil record is usually far from complete and as such it is not possible to guarantee that a fossil taxon is a true ancestor to an extant taxon

(Wiley, 1981). Indeed, it may be far more likely to be just a specialized, extinct side shoot, albeit one of considerable age. Consequently, there is a real possibility that reference to fossils will lead to mistakes in character polarization (Eldredge and Cracraft, 1980). The simplest assumption relating use of fossil data in character polarization is therefore the hopeful one that the oldest known fossils of a group will show a higher proportion of primitive character states. A potentially better way of utilizing fossils comes if there is sufficient evidence to consider that a fossil taxon is an outgroup rather than an ingroup. If a fossil demonstrably represents a 'reasonable' outgroup then its preserved features may be included in outgroup analysis. A further use of fossil evidence comes from their occasional ability to reveal stages in transition series that have been superseded and so are no longer displayed by extant taxa (Novacek, 1992).

A few other criteria for character polarization are also used from time to time and warrant some consideration. The ontogeny of a character has quite often been employed on the assumption that evolutionary advances will tend preferentially to affect the later stages in the development of an organism. Therefore, if a homologous structure present in early development shows a different character state to its later expression in some taxa, then the earlier state is assumed to represent more closely the primitive character state. Essential to this argument is that advanced characters result from the acquisition of extra genes and that those responsible for producing the primitive condition have not been lost from the genome but are simply only expressed (or expressed in an unmodified way) earlier in development (this can result from either von Baer's Law or from recapitulation, the outcome is the same). Nelson (1978) reassessed von Baer's Law from a cladistic point of view and importantly distinguished between recapitulation of a whole organism's form and the developmental sequence experienced by a particular character. The ontogeny argument (Nelson's rule) ignores the possibility that an early stage may itself be adaptive and therefore possess its own derived features. An interesting empirical comparison of the effects of employing an ontogenetic polarization argument and the outgroup comparison method has recently been made by Wheeler (1990b) who applied both separately to larval characters in the leiodid beetle genus *Agathidium*. The result on this occasion was that both approaches appeared almost equally good as estimators of polarity.

Related concepts underlie arguments concerning both vestigial organs, for example, the human appendix or the wings of an ostrich, and occasional minor developmental abnormalities (Scadding, 1981). These two arguments again require that genes coding for primitive character states are not lost but merely have their expression partially or normally suppressed, respectively. In the case of vestigial organs there is the requirement that it is possible to exclude functional from non-functioning states; sometimes this might seem fairly obvious, in other cases considerable work might be required and Scadding has argued that it can never be certain that a structure is without function. The use of developmental abnormalities (teratologies) to determine polarity has been subject to considerable debate, and whilst there can be little doubt that it works in some circumstances, there are many situations in which it is more a matter of faith rather than of science. All of these 'fringe' criteria need cautious treatment and each particular situation should be considered in the context of all other relevant available information, and no doubt arguments will continue for a good while yet (Stuessy and Crisci, 1984).

2.2.3 Homoplasy: convergence, parallelisms and reversals

If all character state changes were truly indicative of evolutionary relationships there would be very little difficulty in reconstructing phylogenies. The problem arises because some character state distributions among taxa result from either parallel or convergent evolution or from the reversal of a character state from the apomorphous state back to its plesiomorphous condition. Collectively, these evolutionary changes which lead to misleading conclusions about phylogeny are termed **homoplasies**.

2.2.4 Homology versus analogy

A fundamental aspect of deciding whether the possession of a derived character state by two different organisms indicates a relationship between them is whether or not these character states have had a single evolutionary origin. If they do represent the same evolutionary event then they indicate that their closest common ancestor is no more ancient than that event. On the other hand, if they evolved independently and merely appear similar they provide no information about the relationships between the two taxa. Advanced states that had a common origin are said to be **homologous**, whereas if they had separate origins and merely appear similar due to parallel or convergent evolution, they are referred to as **analogous**. Whilst the definition of homology is essentially a causal one, that is it is the evolutionary history of two characters that defines whether or not they are homologous, the problem faced by taxonomists is that they do not know the history of the characters and have to be able to infer homology or analogy from other evidence.

Not surprisingly, given the importance of deciding whether similar character states in two organisms are homologous or simply analogous, there has been a great deal written about how such distinctions should be made. One of the most thorough treatments that still forms the basis for most of these decisions was provided by Remane (1956). One possible solution involves the search for intermediates. This predictive approach is based on the notion that if two structures, or for that matter molecular sequences, are homologous then other organisms sharing the same common ancestor as those displaying the postulated homologues should also be expected to show the same feature, and most likely some of these will show intermediate states such that a **transformation series** might be observed (Rieger and Tyler, 1979). This method is sometimes known as the serial criterion.

Another important criterion is that homologous characters should bear the same general relationships to other organs in the organism both in their position and in their developmental sequence. Thus two character states may appear similar in their final fully-developed form but investigation of their development may reveal that they have substantially different origins, potentially excluding the possibility that they are homologous. However, the converse, that two characters may be considered homologous if their position and development is sufficiently similar (Patterson, 1982), leaves open the problem of what constitutes sufficient similarity.

The third most commonly employed criterion is that sets of homologous characters are likely to co-occur in organisms. In other words members of a monophyletic group are likely to display apomorphous character states for several characters. Thus if the two organisms which are suspected of sharing one homologous character state are known with some certainty to share other derived states (homologous characters) then it seems likely that the first characters are also homologous. However, while this argument no doubt works on average, its application in cladistic analysis could potentially involve some circularity and should be treated with care. Further, parallel development of characters is especially likely to occur among closely related organisms which occupy similar niches, and this increases the risk of identifying as homologous characters that are only similar due to parallelism.

2.2.5   *Character state transitions*

When a character displays only two states within the ingroup, i.e. one plesiomorphic and one apomorphic, it is clear that only two types of character state transition can take place, viz. plesiomorphic to apomorphic or vice versa (character state reversal). However, there are numerous situations in which more than two character states occur (see Figure 2.4 a, b, e, f). In these cases more thought has to be given to assessing the possible transitions that may occur since different evolutionary trees are likely to be supported depending on the permitted transitions among the states of a polystate character.

In some situations, variation in a character, perhaps coupled with some common sense intuition, may suggest that a transition series is present, i.e. through the course of evolution a character has changed from one state through a range of intermediates to a new final state, with several of these stages being displayed by the collection of taxa under consideration. For example, in moths it might seem reasonable to propose that taxa with short non-functional wings represent an intermediate state in wing reduction between those of normal fully flighted species and those with wings totally absent. In these cases, the characters should be treated as **ordered** or **Wagner characters**, which means that in calculating the length of the tree for the character, a transition between the extreme states should add a length to the tree equal to the sum of the lengths of transitions via each recognized intermediate state (Figure 2.4 b, e, f; Figure 2.5).

Alternatively, it may not be possible to ascertain what transitions could have occurred or all possible transitions may seem equally likely (Figure 2.4 a; Figure 2.6). In these cases, they would normally be treated as **unordered** (sometimes referred to as **Fitch**) characters. With this option, transitions between any pair of character states are given equal weight and so contribute an equal amount to tree length.

Another important factor that needs to be considered is whether or not a character state transition is equally likely to occur in both directions, i.e. should its contribution to tree length be symmetric.

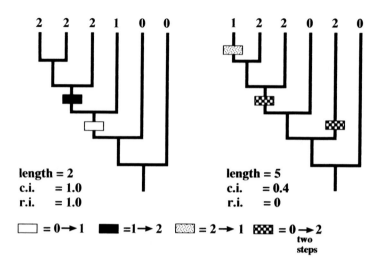

Figure 2.5 Two possible trees showing the distribution of a three state ordered (also called Wagner or additive) character among six extant taxa, together with possible reconstructions of character state transitions that give them minimum lengths (within the constraints of topology and state distributions). Also shown are consistency and retention index values (c.i. and r.i., respectively) which are explained in chapter 3.

an example of the former we might consider the expression of certain chemicals by an organism. For many complex chemicals such as secondary plant metabolites (see chapter 8) a whole series of functional enzymes are likely to be necessary for their synthesis and hence phenotypic expression, not only the one leading to the final product but also all those responsible for producing all the intermediate chemicals in the synthetic pathway. Thus for an organism to display this character a number of genetic states must exist, but for an organism to lose the ability to synthesize the chemical, any one of many possible mutations leading to the loss of activity in any of the intermediary enzymes will suffice. Thus loss of expression of the chemical will be a far more probable event than its independent acquisition. Another example frequently considered in this category is the presence in a genome of sites that can be cleaved by a given **restriction enzyme** (see section 11.4; Jansen *et al.*, 1990). For a restriction enzyme to cut a piece of DNA, an exact nucleotide sequence must be present, but once this sequence has evolved (i.e. the

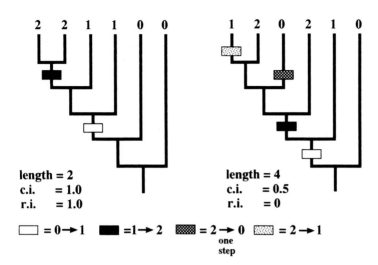

Figure 2.6 Two possible trees showing the distribution of a three state, unordered character among six extant taxa, together with possible reconstructions of character state transitions that give them minimum lengths (within the constraints of topology and state distributions). All possible character state transitions for unordered characters are deemed equally likely to occur and add one to tree length. Also shown are consistency and retention index values (c.i. and r.i., respectively) which are explained in chapter 3.

character state is present), any mutation in the sequence will effectively eliminate the site. Because of the difficulty in exactly quantifying the relative probabilities of the state transitions in each of these particular cases it is simpler to treat such characters as uniquely derivable, more usually termed a **Dollo character**. In reality there is a finite probability that they could arise independently more than once.

An alternative possibility to a Dollo character is a character state which once it has evolved may not be able to undergo any reversal, i.e. character state loss is impossible (Figure 2.4c). Examples of such characters are probably rather limited but some people cite within this category certain chromosomal rearrangements where a subsequent mutation is much more likely to yield a third character state than a state that could be confused with the original one (but see below), or the evolution of polyploidy, although as pointed out by Sivarajan (1991) polyploidy is not necessarily irreversible. Thus, if only two character states are known and one can be assigned as plesiomorphic

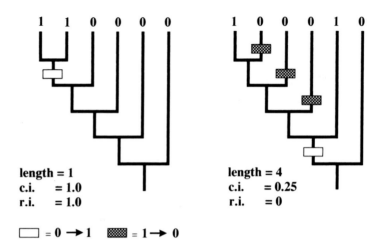

Figure 2.7 Two possible trees showing the distribution of a two state, Dollo character among six extant taxa, together with possible reconstructions of character state transitions that give them minimum lengths (within the constraints of topology and state distributions). Dollo characters are only allowed to evolve to the apomorphous state once though any number of reversals are permitted. The left-hand tree has one acquisition of the apomorphous character state ('white' or state 1) and no reversals. The right-hand tree similarly has only one acquisition of state 1 but has to have three character reversals giving it a total length of 4. Also shown are consistency and retention index values (c.i. and r.i., respectively) which are explained in chapter 3.

(on the basis of outgroup comparison) then it could be reasonably argued that it is highly improbable that the character could have undergone a reversal. Such characters are usually referred to as either **irreversible** or **Camin–Sokal** characters since these latter authors included this criterion in their early parsimon-based algorithm (Camin and Sokal, 1965).

Another possibility, similar in some ways to Dollo characters are polymorphic characters (Figure 2.4 g; Figure 2.9). When a mutation first arises in a population it exists as a polymorphism. Although selection pressures or founder effects may cause such polymorphisms to go rapidly to fixation, i.e. one character state becomes eliminated, it is possible that a character might maintain the same polymorphism after a speciation event. For example, if a population which is polymorphic at a given gene locus becomes divided into two populations which subsequently diverge and ultimately speciate, each of the daughter populations or species might still display the same

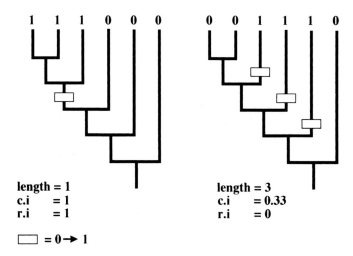

Figure 2.8 Two possible trees showing the distribution of a two state, Camin–Sokal (irreversible) character among six extant taxa, together with possible reconstructions of character state transitions that give them minimum lengths (within the constraints of topology and state distributions). Irreversible characters are permitted to evolve to the apomorphous state on any number of occasions but never to evolve back to the plesiomorphous state. Also shown are consistency and retention index values (c.i. and r.i., respectively) which are explained in chapter 3.

range of alleles. Such characters typically include certain chromosomal modifications such as interchanges or inversions (see chapter 7; Farris, 1978), enzyme polymorphisms (allozymes; see chapter 10), DNA sequences or restriction sites. Probably, however, the number of characters that ought to be treated in this way is considerably underestimated due to failure to study enough individuals of the taxa under investigation and also a failure to be aware of how long polymorphisms may persist in a population. Recent work by Howard (1988) has shown that the latter can be considerable, and in their particular example involving a major histocompatibility complex in mice, the polymorphism appears to have survived at least 3 to 4 million years. As allozyme analyses are often carried out as part of other population genetic studies, workers in this field are perhaps more likely to be aware of intrapopulation variation.

Most extreme, of course, would be characters states that are allowed to evolve only once and then never to be lost, although it is doubtful that such characters could ever occur. The above definition is

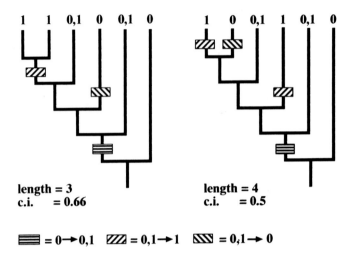

Figure 2.9 Two possible trees showing the distribution of a polymorphic character among six extant taxa, together with possible reconstructions of character state transitions that give them minimum lengths (within the constraints of topology and state distributions). Also shown is the consistency index (c.i.) which is explained in chapter 3.

effectively the same as according the character infinite weight in a phylogenetic analysis that would force those taxa possessing the derived state exclusively to comprise a monophyletic group. In fact of the character types described above, this is the only one which can exclude particular tree topologies; all the others can be accommodated on any tree although at the potential expense of increased tree length.

It should be noted that in addition to the above more or less standard character types, it is also possible for workers to define their own sets of character transformations, assuming that they feel confident enough about the character state transitions that must have occurred (Mickevich, 1982). Typical examples might include various morphological changes or the complex hypothesized chemical pathways leading to diverse secondary plant substances such as the sesquiterpene lactones studied by Seaman and Funk (1983). Of course, in employing user-defined characters, the worker is making a considerable number of *a priori* assumptions that could profoundly influence tree topology and there is always a risk of subconscious bias. Several computer programs used for parsimony analysis allow such user-defined characters but an inevitable consequence of their use is a considerable increase in computing time required to complete an analysis.

The effects of ordered (Wagner), unordered (Fitch), 'evolve only once' (Dollo), irreversible (Camin–Sokal) and polymorphic character parsimony on tree length are shown in Figures 2.5 to 2.9. The significance of the consistency and retention indices (c.i. and r.i., respectively) which reflect how well a character state distribution fits a given tree are discussed in section 3.2.5.

2.2.6   Dealing with missing data and polymorphic characters

It is seldom that every taxon in a data set can be scored for every character. Sooner or later anyone constructing data matrices for phylogenetic analysis will come across situations in which some taxa are either polymorphic for a character or for which the character simply cannot be scored. In the latter case, this may be due to any of several reasons. The organism may be so transformed that a character is not relevant (i.e. eye colour cannot be scored if the organism in question completely lacks eyes), or because the character states were not known, or they could not be observed (e.g. if material has been inappropriately preserved for biochemical tests or insufficient specimens were available to permit dissection or electron microscopy or perhaps the larval stages have not been collected or identified) or more than one character state may be present in the taxon.

The usual way of dealing with both missing values and intrataxon polymorphism is to code the character as unknown, and thus during analysis its state is not defined and does not influence the outcome of the analysis. More precisely, this means that the taxon with the missing character state can be placed adjacent to any other taxon without that character affecting the length of the resulting tree. While this may be the best option for missing data, to code a character that is polymorphic within a taxon as missing can cause problems with parsimony analysis (Nixon and Davis, 1991; see chapter 3). One option may be to try and interpret what the ancestral condition is most likely to have been in the case of each taxon (using independent evidence). An alternative solution becomes apparent when the potential source of the error is considered. The use of a missing value code hides the presence of character state transitions. Because these transitions are ignored in the analysis, the length of the shortest tree(s) constructed from the data will be underestimated, and more importantly, the most parsimonious trees obtained treating polymorphic characters as missing may not be the shortest if instead the polymorphic taxon had been entered in the analysis twice, to take into account separately those members that display each character state.

It is worth noting that missing data are a common problem when trying to incorporate fossil taxa in analyses together with extant ones. Donoghue *et al.* (1989) have considered this problem and concluded that although a data set for a taxon may be incomplete, it should still be included in the analysis even though this may lead to far less resolution in the resulting cladogram. After all, what good is obtaining a more resolved cladogram by ignoring poorly known taxa if that cladogram cannot be relied on.

2.3 Classes of characters requiring special consideration

Many different sorts of characters can be used in phylogenetic studies, e.g. morphological, biochemical, behavioural, macromolecular sequences, etc. These differ from one another in both detail and in more general ways. In particular, differences are evident in the discreteness of their character states, in their reversibility, their reliability and many other things. A few of the classes of character that may warrant particular attention due to these and other factors are discussed in more detail below.

2.3.1 *Characters subject to strong selection pressures*

Several authors have paid considerable attention to the use in phylogenetic analyses of characters that are (or are suspected of being) subject to strong selection pressures because under such circumstances all forms of homoplasy might be expected to be particularly rife. One character set that is cited particularly often in this context is sensory receptors, such as the multitude of cuticular sense organs found in arthropods. Here the worry is that because sense organs are crucial for the interaction of an organism with its environment, it seems highly probable that they will be subject to particularly strong selection pressures.

The possibility that the myriad of small amino acid sequence changes found in many proteins following the introduction of enzyme electrophoresis (see chapter 10) made many consider that much, perhaps nearly all, of the variation must be selectively neutral (Kimura, 1983). This paved the way for a wealth of studies on enzyme variation at all levels from population genetics to species level taxonomy as well as work trying to prove or deny the selective status of the observed variation. As with most other characters it is now realized that these variations comprise a mixture with some under strong selection and others more or less neutral. Nowadays, even though enzyme electrophoresis is far from dead as a taxonomic tool (see for example, Jelnes, 1986), the focus of the search for selective neutrality has shifted more towards DNA sequences themselves. Here large numbers of sites can be found where changes in the nucleotide sequence will have truly negligible effects on fitness.

2.3.2 Environmental effects

Unfortunately from the point of view of the taxonomist, the development of many characters is not an immutable process and the final expression of a character may be influenced by many non-genetic factors such as the state of an organism's health, its nutritional status, its diet, its ambient temperature, its age, its stage in its breeding cycle, etc. How can a taxonomist distinguish between environmentally induced character states and genetically controlled ones? The truth is that all too often taxonomists have failed to do so with the result that non-genetic factors have led to the descriptions of numerous species and subspecies of no taxonomic value. Solutions to this problem depend on taxonomists being aware of the potentially confusing effects of non-genetic variation in the first place, followed by their making an effort to collect and examine as much available data on their specimens and to get to know as much as possible about their organisms' biologies.

2.3.3 Molecular sequence characters

We are in an age when molecular sequence data are becoming available at a phenomenal rate and the taxonomist is therefore challenged to develop appropriate procedures for dealing with this wealth of new information, homology or not. Molecular sequences offer a wealth of characters for phylogenetic analysis given that each position in the sequence can usually be assumed to evolve independently. Thus even a single protein might yield tens or hundreds of amino acid characters while it would require immense effort to find an equivalent number of informative morphological characters for a relatively closely related set of taxa. An obstacle to the use of molecular characters is that they can be expensive to study though costs are now mitigated by the reliability of many off-the-shelf kits. Another initial problem with molecular sequencing that is now being rapidly circumvented was the need for fresh material. Relatively new techniques such as the polymerase chain reaction and gene cloning (see chapter 10) mean that usable quantities of DNA can even be obtained from small parts of long-dead museum specimens. Thus on the face of it molecular sequencing is a very attractive way of uncovering relationships.

However, unlike most other characters, molecular sequence data, i.e. nucleotides in DNA or RNA or amino acid residues in proteins, present a virtually unique problem for the taxonomist. That is, unlike other character systems, it is truly impossible to distinguish homology from analogy for the individual positions in a molecular sequence. Thus, for an amino acid sequence each position can only have one of twenty character states while in a nucleic acid there are only four possible character states for each position. Because of this, homoplasy is likely to be a far greater problem than in the analysis of a suite of morphocharacters. A corollary of this is that one of the basic assumptions of parsimony analysis, namely that evolutionary change is intrinsically unlikely, is often not the case when molecular sequences are considered. Some sequences are conservative in their evolution while others are far more variable. Thus choice of sequence may have a profound influence on the best choice of analytical method.

One particular worry that has originated in many cladistic investigations of sequence data is that they often lead to hypotheses that conflict with those derived from morphological studies. As a group of organisms can have had only one ancestry, such incongruent results mean that either morphology or molecules are giving the wrong answer, or perhaps both are. Hillis (1987) pointed out that congruence between independent character sets provides strong evidence for the given phylogenetic hypothesis, but incongruence is no help. Different molecular sequence data sets can also be (in fact they usually are) incongruent among themselves to some extent. Thus it is rare for cladograms derived from analysis of two separate sequences to be identical. What should be done about this is not certain, and the usual answer is to collect more data in the hope that in the end the trees obtained by analysis of larger and larger data sets will converge towards the one true one. However, as pointed out by Swofford and Olsen (1990) if the 'error' giving rise to the incongruence is a systematic one (i.e. it is an intrinsic property of the data or the analysis method) and if it is of a sufficiently large magnitude, then it does not matter how much extra data is collected, it still will not give the desired answer.

That molecular data do generally behave differently to morphological characters is indicated in many ways (e.g. Figure 3.6). One possible cause for this is that the evolution of morphological character states that give rise to new groups, for example in a major adaptive radiation, may take place over a comparatively short period of time after which successful morphologies may be largely conserved. However, the same is not likely to be true for many molecular sequences which might well accumulate change continuously subsequent to the original splitting events. If this is so, then the ability of sequence data to reveal histories is likely to depend considerably on the relative time scales of the original cladogenesis events under investigation and also the time since the last significant split.

2.3.4   *Electron microscopy and the use of microcharacters*

The use of simple magnifying aids such as hand lenses and compound microscopes has been part of systematics since its inception. However, two twentieth century advances have revolutionized the ability of taxonomists to gain new previously inaccessible morphological information. These are the inventions of the **scanning** and **transmission electron microscopes**, **SEM** and **TEM** for short.

Light microscopy is limited in its resolution of fine detail by the wavelength of visible light (effectively about 550 nm) and by the numerical aperture of the lens (which at best is about 1.4). At high magnifications images start to become confused due to the increasing importance of diffraction phenomena and the best resolution that can be obtained is of the order of 240 μm. Electrons on the other hand with their much shorter wavelengths give the electron microscope a far greater resolving power. Two different types of electron microscopy have been important in taxonomy. Scanning electron microscopy, which makes use of the emission of secondary electrons when the specimen is scanned with a fine electron beam, produces an accurate representation of the surface topology of a specimen. The resolution of the SEM is limited by the minimum diameter of the electron beam and in current microscopes is usually about 3–6 nm. Transmission electron microscopy, in contrast, uses differential electron scatter to visualize chemical and density differences through an ultrathin section of a specimen. The resolving power of the TEM, like the light microscope, is limited by wavelength and numerical aperture of the magnetic objective lens, which being much lower than that of a glass lens (about 0.008) limits their maximum resolution to about 0.25 nm.

Of these two techniques, SEM has had by far the larger impact on taxonomy to date though there is a steadily growing use of TEM for comparative studies of cell ultrastructure which can also have a considerable bearing on taxonomic problems. The development of the scanning electron microscope has greatly increased the range of morphological characters that taxonomists can employ. Preparation of specimens for SEM involves drying and usually coating the surface of a specimen with a thin electrically conducting layer usually of gold, palladium, platinum or some other unreactive metal. Suitable subjects for SEM do not even need to possess hard exteriors but by using appropriate fixation procedures, and when necessary by critical point drying to avoid soft structures collapsing, SEM can produce highly informative pictures even of such delicate structures as isolated cells or protists.

Because of the time-consuming nature of transmission electron microscopic preparation, the problems potentially posed by artefacts, the problem of achieving consistency in technique and the need to fix fresh material, transmission electron microscopy has not been particularly widely used in taxonomy to date. However, there can be no doubt that many ultrastructural features display variation in phylogenetic importance and probably appropriate TEM studies will lead to a greatly improved understanding of relationships.

A few electron microscopes are equipped with a so-called environment chamber that enables them to examine specimens that have not been metal plated and even sometimes specimens that are not fully dried. These systems to date have had limited use in taxonomy although they permit moderate magnification micrographs (usually at least 2000×). Their main use is in the examination of specimens which due to their rarity cannot be gold plated. It should be noted, however, that the specimens are always to some extent damaged due to the electron bombardment to which they are subjected.

Some structures in virtually every group of organisms have been particularly successfully studied by TEM from a phylogenetic view point (Iverson and Flood, 1970; Behnke, 1977; Czaja, 1978; Cole, 1979; Walker, 1979; Duckett, 1988). Of note are a growing number of studies on cuticular excretory organs and spermatozoan ultrastructure covering virtually all metazoan groups as epitomized by many publications by Jamieson (1987, 1991). As Tyler (1979) pointed out, electron microscopy has contributed to taxonomy in two ways; first, it has made available information that was completely inaccessible through light microscopy, and second, it has enabled characters that are visible in light microscopy to be illustrated with far greater clarity, this being particularly true of SEM.

Although it is clear that electron microscopy makes available many additional characters for systematics, there are considerable dif- ferences of opinion as to how far these can be validly used for phylogenetic purposes (Rieger and Tyler, 1979). It has been argued that since small characters are on average also simpler characters and by implication are controlled by fewer genes, they are less informative than macrocharacters. Another important consideration is that they are harder to study from a developmental point of view making the developmental criteria of homology harder to apply both practically and interpretationally. Nevertheless, the recognition of homology for microcharacters must rest on essentially the same criteria as for their larger counterparts (see section 2.2.3).

A separate worry about the use of microcharacters in taxonomy concerns particular classes of microcharacter that might be especially likely to show homoplasy. These arguments centre on their small size and consequent reduced complexity which could mean that they are under the control of relatively few genes, and because with lower complexity they display fewer characters that would enable homology to be distinguished from analogy (Rieger and Tyler, 1979). Erdtmann (1954), for example, showed that while the morphology of pollen grains undoubtedly carries a considerable amount of phylogenetic information, there are also plenty of examples in which distantly related plants produced remarkably similar pollen morphologies.

2.3.5   *Colour as a taxonomic character*

Coloration is well known to be highly variable in many groups and its use in taxonomy should always take into account the possibility that it may reflect intraspecific variation or environmental factors. Colour polymorphisms are widespread and often dramatic in many insect species, with many having seasonally distinct forms that are induced by temperature, humidity, daylength, etc. The reason why colour is usually so variable really stems from the fact that for most larger organisms, it is their external appearance that often determines how potential predators will respond to them. Birds, for example, which are major predators of many insects, are almost entirely visual hunters.

Nevertheless, coloration is seldom useless in taxonomy and for many species reliable identifications can often be made using even minute differences in colour pattern. The critical thing is that the taxonomist or naturalist should be as aware as possible of the amount of variation that each individual species displays. This of course requires that different taxa can be recognized independently of their coloration. If this is not the case, i.e. if two or more distinct colour patterns exist but additional morphological or biochemical characters cannot be found to indicate that they represent different species, it is normal practice to treat these colour types as just forms of a single species, or if they represent geographically separated populations, as subspecies. In general, it is often found that colour pattern is more informative than the exact colour hues present (Wiley, 1981), even more so as colours often fade rapidly in preserved specimens.

2.3.6   *Cryptic and internal characters*

It is a widespread joke that traditional taxonomists spend all of their time counting the bristles on flies' legs or some other similar uninspiring enterprise. Unfortunately, there is a bit more of an element of truth in this than many practising taxonomists would care to advertise. However, what often gets overlooked is the reason why some of us feel at times that we have to perform these potentially mind-numbing tasks. The reality is that evolution has not provided every good biological species with a neat and conspicuous name-tag, and therefore in order to be able to discriminate reliably between species it may be necessary to search for the minutest of features. When the distinguishing features are really obscure they tend to be referred to as **cryptic characters** for obvious reasons, and the pairs or groups of species that they separate, **cryptic species**.

Probably because the classical zoology of the nineteenth century included a large element of comparative anatomy, internal organization has had an important role in taxonomy since its early days, although more so for some groups than others. Techniques of dissection have not changed radically for many years, and if anything training in this area has been on the decline in recent times. However, dissection may sometimes be unavoidable for the reliable identification of some animals and for many others the arrangements of internal or concealed organs may provide a wealth of new phylogenetically informative characters to supplement those available from external study. For some groups, such as nematode and nemertean worms, the extreme uniformity of external characters makes internal anatomy the only realistic path to identification or to phylogeny, and the use of male genitalia in many insect groups for separating species is widely known. However, even among the vertebrates, much of phylogenetic interpretation is dependent on detailed internal anatomy including the structure of the heart, the circulatory system and the brain (Kemp, 1988).

Investigations of internal anatomy generally require fresh or specially preserved material and few museum specimens fit the bill. Birds and mammals are normally skinned perhaps with some skeletal remains being retained but until recently soft tissues have seldom been kept. Insects, especially the larger ones, are usually dry-mounted on entomological pins, and so forth. Nevertheless, even in long-dead dry insect specimens it may be possible to obtain some information on internal anatomy by softening and careful dissection. Modern techniques of preservation can make a big difference and one of the most important advances has been the development of critical point drying, a process in which liquid-preserved material can be dehydrated and dried without the solvent evaporating and causing distortion. Specimens dried in this way can preserve many details of soft tissue anatomy that can be revealed by delicate dissection.

2.3.7   *Animal artefacts*

A surprisingly broad range of organisms produce artefacts of one sort or another. These range from homes and shelters such as burrows and caddisfly cases, nests such as those of birds, fish, bees, wasps and termites, and snares such as made by most spiders and some aquatic insect larvae. In some cases these structures are either robust enough to permit collection, or they may be photographed in the field or casts made for future study.

Differences between animal artefacts reflect a combination of differences in the environment, experience of the constructor and/or underlying genetically controlled behaviours. Whilst all of these may be of interest to the taxonomist if the job at hand is the construction of a diagnostic key to artefacts, it is the last that is paramount if the objective is to use the artefacts to gain a better understanding of the phylogeny. Artefacts are in fact often capable of yielding considerable numbers of characters for phylogenetic interpretation as has been beautifully exemplified by the studies of Wenzel (1991) on nest architecture of neotropical social wasps.

2.3.8 *Behavioural characters*

Animal behaviour is plastic, but it also often has a basis that is largely genetically determined. Some behaviours are very plastic and therefore make poor taxonomic characters, others are stereotyped and can provide good characters. Over the years, ethologists have assembled vast amounts of data on the diverse behaviours of a multitude of different organisms but surprisingly little has been put to taxonomic use. Rather, independently derived phylogenetic hypotheses are commonly used to test ideas about the evolution of behaviour. What does this mean? Does it, for example, mean that behavioural characters are so plastic in developmental terms or homoplaseous in evolutionary terms that they are useless for phylogenetic interpretation? Recent work by De Queiroz and Wimberger (1993) suggests otherwise. However, whether or not this is the case, the fact remains that taxonomists have generally been very careful (even suspicious) about including behavioural characters in their analyses.

Perhaps the one major exception to the general dearth of behavioural characters in taxonomy is as characters for the separation of closely related species. Courtship songs and displays have gained widespread usage in decisions concerning the validity of particular species. For example, genetic isolation between subspecies or sibling species of birds and insects has often been shown to be accompanied by behavioural differences in species recognition (Salomon, 1989; d'Winter and Rollenhagen, 1990). Occasionally, other behavioural features such as activity patterns, for example, are employed for much the same purpose, but their useage is often also correlated more or less directly with reproduction. After all, if one species were strictly nocturnal and another strictly diurnal, it would be difficult to see how they could come together to mate. This sort of behavioural differentiation can even be found within species (Pashley *et al.*, 1992).

Whilst it is obvious that characters such as mate attraction and courtship are likely to correlate well with species isolation, there can be little doubt that far more use could be made of other types of behaviour in taxonomy and in particular, in the reconstruction of phylogenies rather than simply as handles for species delimitation. In this respect it may be worth considering whether a taxonomist would worry so much about including neural morphology as a character system even though neural architecture must ultimately have a strong influence on behavioural repertoirs, particularly in simpler animals.

2.4   Taxa and species concepts

For the purposes of systematic study, taxa can be any group of organisms ranging from individuals through species, genera, families, etc. to kingdoms. Of course most people would not consider individuals to be taxa, though it is frequently necessary to deal with individuals in taxonomy especially when there is uncertainty about to which existing higher category (if any) a particular specimen belongs. In an attempt to avoid this seemingly too broad a definition, Sokal and Sneath (1963) proposed the term **operational taxonomic unit (OTU)**, to include any entity that a taxonomist may wish to treat separately, including individuals. Unfortunately, this term has come to be associated primarily with non-phylogenetic (phenetic) aspects of numerical taxonomy and so has been shunned by many cladists. An alternative term, the **evolutionary unit** or **EU**, has failed to achieve usage as a cladistic alternative.

2.4.1   *Phylogenetic groups: monophyly, polyphyly and paraphyly*

According to cladists the only valid group of organisms are ones which include all the descendants of a single common ancestor, and a special term, **monophyletic**, is used to describe these (e.g. Mayr, 1942; Hennig, 1966; Figure 2.10). However, because much of taxonomy is

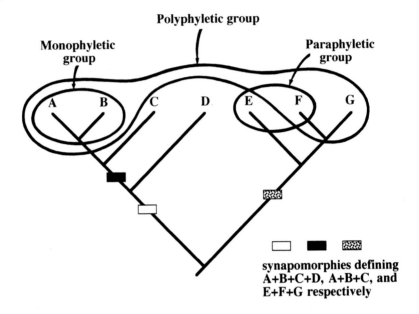

Figure 2.10 Simplified illustration of monophyletic, polyphyletic and paraphyletic taxa as defined by Farris (1974). Monophyletic groups (e.g. A+B, A+B+C, F+G, etc.) include a common ancestor and all of its descendants; paraphyletic groups (e.g. E+F, B+C, C+D, etc.) include an ancestor and some of its descendants; polyphyletic groups do not include the common ancestor of all of their members.

the result of work carried out before Hennigian cladistics came to the fore, many named groups are not monophyletic though they might all have been defined by the possession of some character or other. The precise definition of monophyly has aroused considerable discussion as, for example, Mayr's definition does not explicitly state that a monophyletic group must contain all the descendants of the common ancestor, a criterion that Hennig and more recently Nelson (1971) considered vital. Ashlock (1971) assumed a more Mayrian-type definition of monophyletic and employed an additional term, **holophyletic**, unambiguously to define a group which includes a common ancestor and all of its descendants.

   In addition to monophyletic taxa, two distinct types of non-monophyletic groups are usually recognized, namely **paraphyletic** and **polyphyletic** groups. In the past, however, there has been much debate about the actual difference between these two types (Hennig, 1966; Ashlock, 1971; Nelson, 1971; Farris, 1974; Wiley, 1981). Paraphyletic groups are at first sight the simplest to deal with and are usually defined as comprising some but not all of the descendants of a common ancestor as illustrated in Figure 2.10, i.e. they are monophyletic *sensu* Ashlock, a seemingly simple criterion. The problem arises because members of any polyphyletic group have, at least at some level, a common ancestor, for example, even animals and plants must have had a common origin from a primitive eukaryote. Why then do people want to distinguish two forms of non-monophyletic taxon? Basically because they intuitively feel that they are dealing with two different concepts. For example, a group that would be considered paraphyletic comprising the reptiles which have a common ancestor but do not include mammals or birds seems fundamentally different from a polyphyletic group comprising birds and bats. Paraphyletic groups would therefore seem to give more information about group characteristics than polyphyletic ones and polyphyletic groups carry with them a general notion of having two or more common ancestors. Unfortunately, defining these groups so as to reflect what people feel has proved rather more difficult.

Some of the definitions proposed for para- and polyphyletic groups rely on character state distributions, some on taxa or even on specified numbers of taxa. Hennig's original definitions of para- and polyphyly depended on whether the groups were defined by symplesiomorphies or convergently developed apomorphies (homoplasies). In practice, some taxa could easily be defined by a combination of the two and thus represent both sorts of groups simultaneously. Nelson separated paraphyletic taxa from polyphyletic ones simply on the basis of whether one, or more than one, monophyletic groups of taxa were excluded, respectively. These are clearly exclusive alternatives but ones which could very easily change with the discovery of new taxa or relationships. Wiley, who summarises several sets of definitions, sides with Farris who proposed the following:

Paraphyletic group =a group comprising a common ancestor and some but not all of its descendants.

Polyphyletic group =a group comprising two or more sets of taxa whose most recent common ancestor is not a member of the same group.

Farris's definitions are similar to those of Ashlock and are illustrated in

Figure 2.10. They have the distinct advantage that the two definitions are mutually exclusive and that they collectively include all possible non-monophyletic groups. For the sake of consistency and so as to minimise confusion the same definition is adopted here. However, alternative divisions may also be informative. For example, a group akin to Farris's polyphyletic group could be defined as one which comprises two or more paraphyletic groups. Alternatively, groups may be defined hierarchically so that, for example, the definition of paraphyly could be extended to include traditional paraphyly with polyphyly being a special case. Certainly it would seem sensible that any future new definitions that might be required should employ new terms rather than simply redefining para- and polyphyly which are now becoming relatively settled.

Finally, Donoghue introduced yet another concept which is quite useful, that of the **metaspecies** or **metataxon**. The notion here is that cladistic analyses will often show that the origin of one or more monophyletic group (Figure 2.11) is unresolved with respect to another set of taxa (i.e. the relationships of the other taxa are not fully or unambiguously specified). In these situations some authors may refer to such a group as paraphyletic, although in reality the ambiguity (lack of resolution) in their relationships may well leave open the possibility that they could still form a monophyletic group (Figure 2.11 *lower right*). This ambiguous situation is clearly different from one in which the group is known to be paraphyletic (Figure 2.11 *lower left*) and therefore to distinguish between them is more informative.

2.5   What is a species?

To the irritation of many taxonomists, there is an oft repeated saying that a species is that which a competent taxonomist says is a species. Behind this tautology lie two important considerations. First, there is the problem of what people want a species to represent and second there is the problem of whether an objective definition can be formulated that will satisfy the answer to the first problem. Together these constitute what may be termed the species definition problem.

One fundamental but often overlooked aspect of the above is that species definitions can be framed either in terms of direct attributes of the individuals constituting a species, or they can be couched in terms

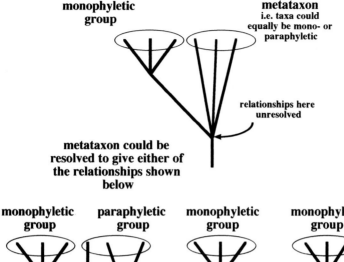

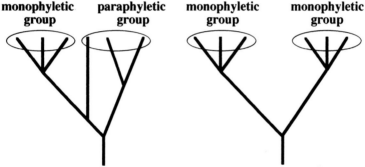

Figure 2.11 An illustration of why a metataxon is not necessarily paraphyletic. Metataxon relationships are not fully resolved and therefore could allow the possibility of still being monophyletic.

of a process that, by definition, gives rise to species (Donoghue, 1985; Lidén and Oxelman, 1989). Within both frameworks there are several competing definitions and some definitions even combine aspects of both process and attribute. Whilst both types of definition have their own merits, in practice only the former can act as benchmarks for the practising taxonomist although the user of taxonomy may have conflicting wishes.

Over the past century or so, many authors have proposed their own definitions of what a species is, far too many to review here. Several discussions and suggestions are included in a recent book edited by Ereshefsky (1992), but in essence they fall into three categories: the widely referred to 'biological species concept' based on reproductive isolation; phylogenetic and evolutionary species concepts based on basic evolutionary units as revealed by phylogenetic analysis; and *ad hoc* definitions which might be broadly classed as morphological species criteria based on degrees of phenotypic or genotypic distinctiveness. These groupings cut across the distinctions proposed by Lidén and Oxelman. For example, the biological species concept (section 2.5.1) can be framed in terms of either process or attribute.

2.5.1 *Biological species concept*

Many zoologists hold the notion that a species can be defined as a group of individuals which form an interbreeding population such that gene flow can occur freely within a species but hardly ever occurs between species. Several definitions of what a species is have been based on this notion, the most widely known of which was proposed almost exactly 50 years ago by Mayr (1942). His view is now widely referred to as the **biological species concept** and is roughly summarized as follows:

> A species is a group of interbreeding populations that are genetically isolated from other groups by reproductive isolating mechanisms such as hybrid sterility or mate acceptability.

This definition although quite widely accepted among evolutionists, has nevertheless been the target of much criticism particularly from systematists (e.g. Sokal and Crovello, 1970; Cracraft, 1983; Donoghue, 1985) and equally vigorously defended (Mayr and Short, 1970; Mayr, 1992). The problem might be more or less resolved if it were accepted that no one definition will suffice equally satisfactorily for all organisms and perhaps not even for all purposes. Acceptance of this, however, is intuitively less tidy than the idea that a single definition can suffice for everything. More importantly, it should be asked why many evolutionary biologists should seemingly be at odds with systematists.

The central issue seems to concern whether species are viewed as the units of evolution or as potentially inbreeding sets of populations which might therefore display some coherence from a population biological point of view. The two views about what constitutes a species (or should do), are not the same and do not necessarily lead to drawing the same boundaries. There has even been considerable

debate about what constitutes the basic unit of taxonomic differentiation — individuals, populations, subspecies, species, etc.

Probably the strongest arguments against the biological species
concept come from phylogeneticists. For example, Donoghue (1985)
noted that the biological species concept does not necessarily yield
monophyletic species (see section 2.5.4), whilst Baum (1992) pointed
out that relatedness and ability to reproduce are not so tightly linked
as many would assume and that this aspect may also vary between
major taxa. Further, many botanists, because of the general ease with
which plants can hybridize, have specifically rejected the biological
species concept. Plants differ from animals in this respect for several
reasons. These include differences in the complexity of their
organization such that genetic imbalances that would be fatal to many
hybrid animal embryos are less likely to prevent hybrid plants from
developing to maturity. Most plants are hermaphrodites and once
mature many hybrid plants can carry on trying to reproduce for a long
period of time such that the chance of successful self-fertilization or
backcrossing is greater than for many animal species. Collectively,
these and other biological features of plants seem, at least at first sight,
to render the biological species concept almost entirely impracticable
for them. However, that view has recently been challenged by Mayr
(1992) who critically appraised how well the biological species concept
could be applied to an entire local flora in North America. Perhaps
surprisingly, he concluded that it was perfectly applicable to some
93.5% of taxa.

Of course, the biological species concept has no relevance to
parthenogenetic species, for example plants that habitually undergo
agamospermy such as the dandelions (*Taraxacum*), and neither is it
relevant to asexual taxa, and so among other things it cannot be
applied to many Protista or Prokaryota. That is not to say, however,
that among these latter groups genetic exchange between individuals is
non-existent. Indeed there is growing evidence that many bacteria and
Protista regularly conjugate to exchange genetic information. This has
in turn lead to the concept of the **genospecies** which is defined as a
group of bacterial strains, varieties, etc. that are readily capable of
exchanging genetic information.

Even among zoologists, the notion that species can be defined solely
by their ability to hybridize has received widely differing reviews.
Dubois (1988), following an extensive survey of vertebrates, has
proposed that the ability to form viable hybrids is more useful as a

criterion for inclusion of species in a genus rather than as an argument for inclusion of the taxa in the same species. In an effort to overcome this, Geist (1992) proposes that rather than simply considering whether a mating between two species can produce a viable offspring in captivity, the question should be asked whether such hybrids would be able to survive in the wild. Thus, whilst a hybrid between two animals might grow into a healthy individual in captivity, for example the cross between the American white-tailed and mule deers (*Odocoileus virginianus* and *O. hemionus*, respectively), these individuals fare badly in the wild when exposed to natural predators because they display a defence strategy which, being intermediate between those of their parent species, is maladapted. Hybrid animals often show behaviours and morphologies intermediate between those of their parents, and thus if the parental species have distinctive courtship behaviours and associated structural and chemical signals, their hybrid offspring are unlikely to be particularly successful, if at all, in attracting a mate. Of course, under artificial conditions as found in many zoos, the enforced close proximity between species will quite often result in hybridization. In the wild, such hybrids will probably be rare and the chance of them mating with anything ever rarer. Such secondary mechanisms are by no means restricted to animals. An interesting type of barrier that can limit or prevent fertilization of a hybrid plant in the wild is pollinator maladaptation. Many species of insect-pollinated plants in the wild are adapted to pollination by a small group, or even by a single species of insect (O'Toole and Raw, 1991). For example, crosses can be made between different species of a plant whose flowers are each adapted to pollination by a different sort of organism. The flowers of the F1 hybrid will, however, probably be maladapted and under natural conditions may not succeed in attracting any suitable pollinator because they show a mixture of those factors required. Unfortunately, while the sorts of investigation necessary to discern such secondary barriers to introgression may be carried out for a few species, it is impracticable for the great majority even of higher vertebrates and plants yet alone the multitude of invertebrates, etc.

Another problem of the biological species concept is that it can be applied to **sympatric** populations with relative ease but this is not so with **allopatric** ones (Cracraft, 1982). In the former case it is usually possible to determine whether putative species mate together and, for example, by means of **allozyme** studies (see chapter 10), to assess how much gene flow is occurring. However, with putative allopatric species, the systematist has a larger set of problems. First of all, being allopatric, there cannot be any significant gene flow and therefore the systematist is effectively left having to guess whether the two would interbreed if they were sympatric. Simply bringing the two into captivity or cultivation and seeing if they will cross is not sufficient as it is well known that under such artificial conditions many species will cross which would not normally do so in the wild even if they were sympatric. Further, if a closely related pair of sympatric species in the wild were occasionally to produce hybrid offspring with reduced inclusive fitness, then natural selection will tend to reinforce barriers to mating, such as species-specific courtship rituals, whereas with allopatric species such isolating mechanisms are likely to evolve far more slowly.

2.5.2   *Phylogenetic species concept*

The above criticisms of the biological species concept of necessity led to several alternatives that do not rely on knowledge of breeding potential (Wiley, 1981). Of these, probably the most relevant to the needs of the practising taxonomist is the **phylogenetic species concept** of Cracraft (1983) and its many subsequent modifications. The main difference between these and the biological species concept is that they do not consider present biological characteristics directly but rather the acquisition of defining features during evolution (Donoghue, 1985; Baum, 1992). Thus, one widely held definition runs as follows:

> A species is the smallest possible group of a sexually reproducing organism that possesses at least one diagnostic character which is present in all group members but is absent from all close relatives of the group.

Diagnostic characters of phylogenetic species can be any genetically determined feature including behavioural ones. The criterion that diagnostic characters are fixed in the population together with the criterion that the population is a sexually reproducing one rules out the possibility that males and females could be classified as separated species since neither males nor females in ordinary sexually reproducing species can procreate by themselves.

The above phylogenetic species definition, however, is not without its drawbacks, one of which being that the criterion of diagnosability allows for the use of plesiomorphies as well as autapomorphies and therefore any species thus defined by a plesiomorphy could potentially be paraphyletic. Accordingly, modified definitions have of course been constructed to exclude this possibility by explicitly adding requirements for monophyly to the definition (Donoghue, 1985; Mishler, 1985; De Queiroz and Donoghue, 1988, 1990). As pointed out by Wheeler (1990) application of the phylogenetic species concept would almost certainly give far larger estimates of the total number of species than the more traditional biological species concept. This is taken by up Mayr (1992) who illustrates it by reference to Rosen's (1979) work on South American poecilid fish. Mayr points out that application of a phylogenetic species concept in this case would mean that the distinct populations of many species that inhabit almost every tributary would then have to be raised to species rank. This would seem to entail an unnecessary degree of complication.

2.5.3   *Evolutionary species concept*

The **evolutionary species concept** was first proposed by Simpson (1961) though as with most other species concepts it has undergone a series of revisions and improvements. One variant due to Wiley is probably representative:

> A species is a single lineage of ancestor–descendant populations which maintains its identity from other such lineages and which has its own evolutionary tendencies and historical fate.

This definition while seeming to make sense in that each evolutionary species more or less corresponds to a branch (including internal ones) or a part of a branch in an evolutionary tree, has the practical disadvantage that it is not possible at any one time to know the historical fate of an extant lineage, and therefore this can only be surmized from other types of data. Wiley (1981) discusses this species concept in greater depth than space permits here.

2.5.4   *Problems with parthenogenetic species and asexual clones —*
some further considerations

Unfortunately there are quite a number of species which normally reproduce entirely by parthenogenesis such that males may be unknown and either occur only rarely or not at all. In these cases the biological species concept breaks down because each individual of a parthenogenetic species is reproductively isolated from all others (De Queiroz and Donoghue, 1988, 1990). Parthenogenesis, in fact, presents many species definitions with potential headaches, and a couple of interesting examples will suffice to illustrate the difficulties they can pose.

Many 'species' of lichen (or to be precise, lichen-forming fungus; Tehler, 1982) are known from two forms, one reproducing by normal sexual means and producing spores containing only fungal material, the other reproducing asexually and producing **soredia**, spore-like propagules containing both fungal and algal components. In these cases it seems that after a soridophyte form has evolved from a normal spore producing lichen, there is no return. Thus each asexual clone that results is no longer a part of the same biological species as its original sexual ancestor. They also differ in a distinctive attribute, namely their reproductive mode. However, if each of these soridophyte forms are treated as separate species, they leave their ancestral sexual species paraphyletic!

Another interesting example occurs among various groups of parasitic wasps in which some entirely parthenogenetic all-female 'species' are known together with apparently closely related sexual ones. In total, more than 200 such thelytokous strains are known, collectively representing more than 20 families of wasps. Surprisingly, several of these all-female producing 'species' have been shown to be the result of an infection by an as yet unidentified microorganism. In these treatment with either high temperatures or antibiotics can induce the production of males. In the case of some genera such as *Trichogramma* this is particularly important, as species level taxonomy is often based largely on male characters which would otherwise remain unknown were it not for this ability to 'cure' the species in culture from its presumed infection (Stouthamer *et al.*, 1990). Whether these microorganism-induced asexual strains should be regarded as separate species now that it is known that male production can be achieved artificially is problematical. This is particularly so as it appears that in some cases the males produced as a result of antibiotic or heat treatment are sterile.

CHAPTER THREE

PHYLOGENETIC RECONSTRUCTION — CLADISTICS AND RELATED METHODS

It is crucial in what follows that the reader keep a certain simple distinction clearly in mind. Criticising the logic of an argument is quite different from criticising that argument's conclusion.

Sober (1988)

3.1 Cladistics and cladograms

This chapter is concerned with the current methods that are available for estimating phylogenies primarily through the analysis of character versus taxa data matrices but one section deals with the use of taxon-to-taxon distance measures from a phylogenetic perspective. Three rather different methods based on cladistic principles have been more or less widely employed for this purpose. These are **parsimony analysis, compatibility analysis** and **maximum likelihood analysis**. The first of these has been by far the most widely accepted for a variety of practical and theoretical reasons and will be largely concentrated on in the discussion that follows. However, while the three above-named types of procedure encompass nearly all of current cladistic practice, it should not be imagined that the azimuth of tree-constructing methods has been reached. Many problems still remain. Unequal evolutionary rates along different evolutionary paths and differing degrees of homoplasy in different characters pose their own problems, while for analysis of macromolecular sequences many issues concerning alignment, deletions and insertions, and bias in nucleotide composition have yet to be fully resolved. Yet more difficulties can potentially result from hybridization events and **introgression** (horizontal gene transfer) since all current methods of cladistic analysis presuppose that characters are only inherited in simple ancestor/descendant lines.

Finally, but certainly not the least important, is the shear magnitude of the computing problem involved in finding the best trees for more than a small number of taxa. All of the above beg improved procedures and there is no shortage of research in this area. Penny *et al.* (1992), for example, have briefly summarized several up and coming approaches to phylogenetic reconstruction, i.e. **invariants**, evolutionary parsimony and 'the great deluge algorithm', the latter so-called because of its rapid searching for a large number of diverse trees, though these are not as yet in widespread use and only invariants will be dealt with further here.

3.1.1 *Parsimony*

One of the basic tenets of much of cladistic analysis, but one which has nevertheless been subject to substantial criticism from time to time, is that the most likely explanation of a taxonomic data set is the one which requires the least number of evolutionary changes or, more specifically, character state transitions (Felsenstein, 1978b, 1983; Friday, 1982; Farris, 1983, 1986). Such explanations take the form of trees connecting the taxa under consideration and the shortest trees derived from the data set, i.e. the one(s) that require the fewest character state transitions to have taken place, are known as the most **parsimonious** ones. Methods used to find these most parsimonious solutions are accordingly referred to as **parsimony analysis**.

One of the most important assumptions of parsimony analysis is that character state transitions are intrinsically unlikely events. Otherwise, if they were considered as probable events, it would be a nonsense to try to find evolutionary hypotheses that minimise them. Conversely, if a data set is suspected of being heavily biased towards characters undergoing rapid and homoplaseous change, then parsimony analysis should not be expected to provide accurate phylogenetic estimates.

3.1.2 *Compatibility analysis*

Whereas variation in some characters may truly reflect the evolutionary history of a group (**true characters**), others, due to homoplasy, do not (**false characters**). In practice many characters usually show some homoplasy (see section 2.2.3) and so different characters will often support different phylogenetic hypotheses even though only one hypothesis can be correct. The problem is that frequently there is no totally reliable means of distinguishing between true and false characters. A possible solution to this was suggested by Le Quesne (1964) who argued that: 'Variation in any combination of true character will always be compatible, that is, it will always support the same phylogenetic hypothesis, whereas, if two character state distributions are incompatible, then at least one of the characters must be false.' To minimize the effects of false characters, Le Quesne proposed that analysis should be restricted to characters that are compatible with one another and thus the raw data set should be pruned down so as to leave the largest set of characters that are mutually compatible. In its underlying philosophy this is not very different from parsimony analysis in that both aim to 'minimize' homoplasy (Felsenstein, 1983).

The essence of performing compatibility analysis is to find a set of mutually compatible characters, usually called a **clique**, from among all the available ones. The larger this set is relative to the number of taxa, then the more likely it is on average that relationships among the taxa will be fairly well resolved. The larger the clique relative to the total number of characters, then the more confidence, rightly or wrongly, there may be in the data set. Unfortunately, if the compatibility idea is accepted, perhaps because it intuitively seems to be conservative, the largest clique will quite often be found to contain only a small number of the original characters. For example, Meacham and Estabrook (1985) surveying 22 published data sets found that the proportion of the total number of characters represented in the largest clique ranged from 0.16 to 0.96 with a mean of 0.46. Thus, on average nearly half of the characters scored were discounted in the subsequent tree construction process, and any phylogenetic information they contained would not be used. Many proponents of compatibility analysis see this as a distinct advantage in that if the largest clique comprises only a few characters then the data may not be up to the job of indicating the group's phylogeny. One way of still following a compatibility procedure that in part overcomes the problem of discarding a good proportion of potentially informative data is to keep applying the analysis to the progressively smaller subsets of taxa that were indicated as monophyletic by the previous analysis. Given that there may be a tendency for parallelisms to occur within groups of fairly closely related taxa this may be reasonably acceptable.

Several computer implementations have been provided to perform the role of finding the largest set of compatible characters, for example, CLINCH written by K. L. Fiala and described by Estabrook *et al.* (1977).

3.1.3   *Maximum likelihood and related methods*

While much evidence suggests that parsimony analysis (see section 3.2) does quite a good job of estimating phylogenies, it has been shown that there are circumstances in which it will fail (Figure 3.1; Felsenstein, 1978b). Specifically, parsimony will fail consistently if evolutionary rates along different evolutionary branches are sufficiently dissimilar and if the assumption that change is rare is violated. More reliable estimates of phylogeny can be made if information about evolutionary processes can be included in the analysis. In these models, evolutionary change is assumed to be a purely chance process and as a result each possible phylogeny for a set of taxa must have a certain probability of being correct given a particular set of data. Thus

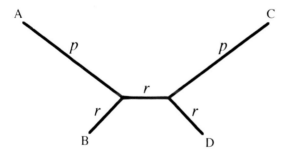

p, r = rates of evolution, $p \gg r$

Figure 3.1 An illustration of unequal evolutionary rates leading to four taxa that under the condition of rate p being much greater than r will cause parsimony analysis to fail to lead to the correct evolutionary interpretation. The reason for this is that with enough changes along the branches leading to A and C, chance will mean that some will give rise to analogous apomorphous states in both taxa. Hence, if $p \gg q$ then the number of apomorphies shared by A and C will exceed those shared by A and B or by C and D. This model is particularly relevant to molecular data where homoplasy is likely to be common and parallel changes truly indistinguishable. (After Felsenstein, 1978.)

the preferred tree should be the one that is most likely to have given rise to that particular data set, and the approach that finds this tree is referred to as a **maximum likelihood method**. Under some circumstances, for example if change is not rare, maximum likelihood will yield quite different trees to maximum parsimony.

At present, and probably for the foreseeable future, maximum likelihood treatments are likely to be largely limited to analyses of a very few molecular characters where the evolutionary transitions comprise a small and tightly defined group of changes whose probabilities can be estimated with some accuracy. Consequently, most applications of maximum likelihood methods have dealt directly with nucleotide sequence data (Swofford and Olsen, 1990), although DeBry and Slade (1985) have also considered its application to the analysis of restriction site data (see section 11.5), Swofford and Berlocher (1987) have applied it to allozyme frequency data (see chapter 10) and Rohlf and Wooten (1988) have tested it to a genetic drift model of allele frequencies. One drawback of maximum likelihood approach is that it is demanding of computer time.

A particularly awkward problem that has dogged phylogenetic analysis using sequence data is that some of the changes that occur along different diverging branches of the evolutionary tree will by chance be parallel. Further, the more changes that take place along two lineages, the greater the number of parallelisms there will be. However, unlike some other sorts of data, there is no way of re-examining molecular data to try and distinguish an homologous change from an analogous one (in this case a parallel change). One possible way of minimizing the effects of parallel change is only to consider relatively conserved sequences, another is only to count **transversions** (relatively rare DNA base changes in which a pyrimidine is replaced by a purine or vice versa; see chapter 11), and to carry out parsimony analysis on these changes — this process has been called **transversion parsimony**. However, the latter results in discarding much potentially useful information. Recently, Lake (1987) has proposed an alternative method which similarly only considers transversions but does so in a different way. Lake's technique, referred to as **Lake's method of invariants** or **evolutionary parsimony**, is restricted to sets of four sequences (taxa) and further considers only base positions in the sequence where two taxa have purine residues and the other two have pyrimidines (Figure 3.2). The idea behind Lake's method is that if this distribution results from a parallel change then it is equally likely that

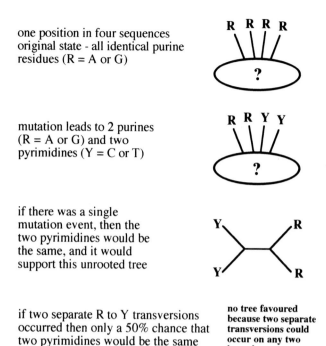

one position in four sequences
original state - all identical purine
residues (R = A or G)

mutation leads to 2 purines
(R = A or G) and two
pyrimidines (Y = C or T)

if there was a single
mutation event, then the
two pyrimidines would be
the same, and it would
support this unrooted tree

if two separate R to Y transversions
occurred then only a 50% chance that
two pyrimidines would be the same

**no tree favoured
because two separate
transversions could
occur on any two
branches**

**if the number of positions with 2 different pyrimidines
(i.e. 50% of parallel events) is subtracted from the
number of positions with 2 identical pyrimidines, the
answer should approximate the number of potentially
informative transversions**

Figure 3.2 Basis of Lake's method of invariants, a technique for inferring evolutionary
relationships for sets of four taxa based on their nucleotide sequences. Abbreviations: A,
adenine; C, cytosine; G, guanine; T, thymine; R, either purine; Y, either purimidine.

the resulting new pair of bases will be identical or different (i.e. two
pyrimidines could equally change to two adenines or two guanines as
they could to an adenine and a guanine). However, if the change was
due to a single informative apomorphic event, then the resulting bases
will always be the same. Using this inequality, Lake's process then
counts which of the possible unrooted four-taxon trees is supported or
is countered by a particular distribution of transversions. Because it

makes a distinction between different purines or pyrimidines, Lake's method uses more information than transversion parsimony (Swofford and Olsen, 1990). However, the value of Lake's method is limited by the relative frequency of **transition** substitutions (DNA base changes in which a pyrimidine or purine is replaced by another pyrimidine or purine, respectively). If transitions are too common, then they will be likely to mask informative changes.

3.2   Parsimony and finding the shortest trees

This section deals primarily with parsimony analysis and different ways of processing and interpreting the trees that result although some of the techniques discussed are also relevant to the trees produced by other methods, particularly those concerned with rooting trees, consensus trees and comparing trees.

3.2.1   *Finding the shortest trees and the impact of computerization*

At first sight the problem of finding the shortest tree diagram that can connect a group of taxa may seem relatively simple. After all, in theory every possible tree for the group of taxa concerned can simply be measured and the shortest (i.e. the most parsimonious one) taken as the preferred cladogram. However, the problem is often far from trivial because of the huge number of different possible trees that could be involved. Section 3.2.2 deals rather more mathematically with the problem for those who are interested, but in essence all that should be realized is that the number of possible trees connecting a set of taxa increases ever more rapidly with the number of taxa involved.

It has been shown that it is highly unlikely that any reasonably quick and practical analytical procedure will ever be found to determine the shortest possible tree or trees for any given set of data and therefore the discovery of the most parsimonious trees must rely on searching strategies. However, searching methods have their own limitations due to the potentially astronomical number of possible trees that may need to be searched (see section 3.2.2). It is obvious therefore that even for analyses involving quite small numbers of taxa it would be extremely time-consuming to evaluate all possible trees by hand and even with the advent of reasonably fast microcomputers evaluation of the lengths of all possible trees (**exhaustive search**) is only really feasible for situations with up to ten or so taxa, a situation that will not change dramatically even if computer speeds should increase a thousand-fold over the next few years.

As taxonomists frequently need to deal with far larger data sets, methods other than brute force exhaustive searches are essential. Thus several highly sophisticated algorithms have now been devised to search for minimal length trees, though with the exception of the **branch-and-bound** method described below, none of these others, which are collectively referred to as **heuristic** methods, can guarantee to find the shortest tree or trees.

Branch-and-bound algorithms provide a simple guaranteed method of finding minimum length trees which will usually greatly reduce computing time and therefore permit exact analysis of larger data sets than would be possible using exhaustive searching (Hendy and Penny, 1982; Platnick, 1987; Swofford, 1991). As with an exhaustive search, these algorithms set out to build step-wise all possible trees; however, instead of evaluating tree length only when all taxa have been added to a tree, its length is calculated at every intermediate stage. If at any stage in the tree-building process the length is found to exceed the current minimum tree length, then all other trees that contain the same arrangement of taxa can be discounted because tree length can never be reduced by the addition of further taxa. The number of taxa that can be handled by the branch-and-bound method is data set-dependent. The method's efficiency also depends on the starting value for the minimum tree length above which partially constructed trees need no longer be considered. In practice, branch-and-bound methods usually allow confident cladistic analysis of up to 20 or 25 taxa. Increased efficiency can be obtained by using an approximation of the minimum length tree based on a preliminary heuristic analysis (Swofford, 1991).

The use of heuristic methods for finding shortest trees comprises two stages: an initial tree building stage followed by a series of rearrangements to improve on the first attempt. In the most simple scenario, calculation of these trees is unrooted, and the root point is determined only after the shortest trees have been obtained. Thus in this case, character polarity need not be known in order to find the most parsimonious tree, and is only needed for the purpose of inserting the root. However, this methodology may be inappropriate for both philosophical and practical reasons. Philosophically, evolution itself starts from an ancestor and so it makes more biological sense to include a hypothetical ancestor or an outgroup in the analysis from the beginning. In fact, doing so can yield shorter trees than will be found by calculating an unrooted tree first and adding the root subsequently. From a practical point of view, if Dollo, partially reversible or irreversible characters are employed in the data set then the explicit asymmetry of gain or loss of apomorphous states of these characters means that parsimony analysis cannot be performed without fixing the ancestral conditions beforehand (but see Swofford and Olsen, 1990).

A number of computer packages are in widespread use for conducting exhaustive branch-and-bound and heuristic searches. Among the most popular and best are versions of PAUP (phylogenetic analysis using parsimony) by Swofford (1991) and Hennig86 by Farris (e.g. version 1.5; Farris, 1988) though a few useful algorithms are also implemented by the PHYLIP programs (phylogeny inference package) of Felsenstein (1987) and the PHYSIS package of Mickevich and Farris. The efficiencies of several of these have been compared by Platnick (1987, 1989) but new and improved versions of programs are appearing all the time. With current increases in speed and memory of personal computers, even quite large data sets can be analysed (at least heuristically) with relative efficiency.

3.2.2   Tree facts and figures

The topological properties of trees impinge in several important ways on character analysis and phylogenetic reconstruction, and therefore a number of the more pertinent considerations are dealt with separately in this section so as not to interrupt the flow of argument in later sections. Those readers who are not particularly interested in the mathematics underlying the number of different ways members of a group of taxa can be related to one another can simply skip this section.

A few terms used here require definition. First, the term **labelled** refers to a tree in which the taxon at the end of each external branch is specified (or specifiable) while **unlabelled** trees are simply topologies without regard to the identities of the taxa they connect. Second, **rooted** trees are ones in which one terminal branch is specified (or is specifiable) as an ancestral taxon, this is not counted among the number of taxa that the tree connects (Figure 3.3).

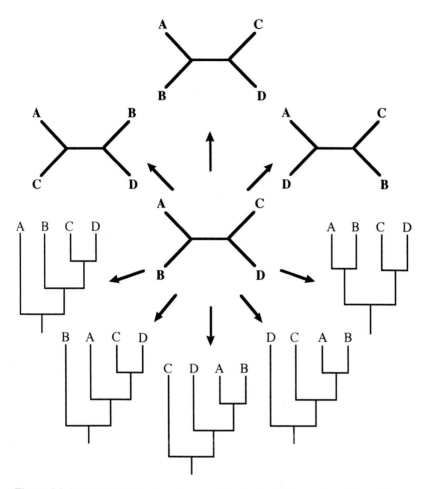

Figure 3.3 Diagram showing how four taxa (A, B, C and D) can form three different unrooted labelled trees (*above*) and fifteen different rooted ones of which five based on the centre figure are illustrated (*below*).

One of the major limiting factors encountered in cladistic analyses is that the number of distinct bifurcating labelled trees increases very rapidly with the number of taxa under consideration. Thus even with modest numbers of groups there may be millions of different possible trees. The easiest way to understand this is to consider the number of ways in which a new taxon can be added to an existing tree. In fact, for any unrooted tree with labelled terminal taxa, adding a new taxon to any of the branches will give a distinct new tree. Therefore, the total number of different ways a new taxon could be added is equal to the number of branches in that tree, which for any existing unrooted tree of t taxa will be $2t - 3$, i.e. t branches leading directly to terminal taxa and $t - 3$ between the common ancestors of the terminal taxa. However, the t in the $2t - 3$ expression refers to the number of taxa in the tree before the addition of the new taxon and therefore in order to find the number of trees obtainable from that starting tree in terms of the new number of taxa n we have to make the substitution $t = n - 1$. This gives the number of labelled trees of n taxa, obtained by adding a single new taxon to a given previous tree as $2n - 5$.

It follows from the above that the total possible number of labelled unrooted bifurcating trees with n taxa will be given by the product of $2n - 5$ and the number of distinct trees with $n - 1$ taxa, and this will in turn depend upon the number of trees with $n - 2$ taxa and so on down to $n = 3$ for which there is only one possible unrooted variant (Felsenstein, 1978a). Thus:

$$\text{No. of trees } (N) = (2n - 5) . (2(n - 1) - 5) \ldots \ldots (2(4) - 5) . (2(3) - 5)$$

$$= \prod_{i=3}^{n} (2i - 5) \qquad \text{or} \qquad \frac{(2n - 5)!}{2^{n-3}(n - 3)!}$$

From the above equation it can be seen that when the number of terminal taxa $n = 3$ there is only one possible arrangement, with $n = 5$ there are 15 arrangements but when $n = 10$ there are over 2 million arrangements and when $n = 50$ there are more than 10^{74} distinct trees.

Taxonomists are of course particularly interested in the number of rooted and labelled trees, that is ones in which the common ancestor has been defined. This number can easily be obtained from the above equation by multiplying by the number of places at which a tree can be rooted. This again is equal to the number of branches, i.e. $2n - 3$. Thus the number of different labelled bifurcating and rooted trees is given by:

$$N = (2n - 3) . \prod_{i=3}^{n} (2i - 5)$$

$$= \prod_{i=3}^{n} (2i - 3) \quad \text{or} \quad \frac{(2n - 3)!}{2^{n-2}(n - 2)!}$$

and this relationship is illustrated in Figure 3.4.

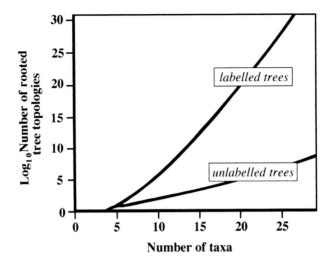

Figure 3.4 The relationship between the number of taxa and the number of different terminally labelled and unlabelled rooted trees that can be formed from them.

In addition to labelled trees, it can also be informative to consider how the number of unlabelled trees (tree topologies) is related to the number of terminal taxa. However, this is not such a straightforward problem as the one above. Indeed, for unrooted trees there appears to be no explicit formula for calculating the number of topologically distinct forms. Harding (1971) has provided an algorithm for calculating the number of different rooted unlabelled tree topologies which is based on the number of permutations in which two smaller trees can be combined to make a tree of the desired number of terminal branches. Thus, it is necessary to know how many distinct topologies there are for trees of all sizes up to $n - 1$ termini. These values can of course be calculated by the same method from values for smaller trees, and therefore it is only necessary to seed the method with the starting values of 1 for $n = 1$ to $n = 3$ and of 2 for $n = 4$. If then the number of topologically distinct rooted trees with n termini is S_n, then

$$S_n = S_1 S_{n-1} + S_2 S_{n-2} + \ldots + S_{(n-1)/2} S_{(n+1)/2}$$

for odd n

and

$$S_n = (S_1 S_{n-1} + S_2 S_{n-2} + \ldots + S_{(n-2)/2} S_{(n+2)/2}) + (S_{n/2}(S_{n/2} + 1))/2$$

for even n. The relationship is approximately logarithmic as can be seen from Figure 3.4, but increases at a considerably lesser rate than the number of different labelled trees.

Finally it should be noted that in all of the above it has been assumed that the taxa are all located at the ends of their own discrete evolutionary branches and are not connected in direct ancestor–descendant relationships. If taxa are allowed to occupy positions at the internal nodes of a tree then the number of possible trees is greatly increased (Platnick, 1977b).

3.2.3 Building trees from distance data

While most data employed in systematics are obtained in the form of character state versus taxon matrices, a few techniques yield data in the form of taxon-to-taxon distances (or similarities). Two procedures are particularly important in this context because of their widespread use in taxonomy, namely microcomplement fixation (MCF) and other immunological procedures (see chapter 9), and DNA hybridization (see chapter 11). Unfortunately, although only a few other techniques yield distance data directly, there has been a tendency in some other areas to convert ordinary taxon versus character data into distance measures and then to use this downgraded information for tree construction. This approach has been particularly prominent in work on allozyme data which was initially employed largely in population genetic studies in which indices such as Nei's genetic distance were routinely employed (Nei, 1972; Rogers, 1972; see chapter 10). In other areas, the same trend was probably more a result of the greater familiarity of many workers with the numerical taxonomic approaches of Sokal and Sneath rather than with cladistic methodology. In addition, the conversion of discrete character data into distance measures prior to analysis was no doubt further encouraged by the greater availability of computerized phenetic clustering techniques compared with parsimony procedures. Views as to whether applications of distance techniques provide any useful information lead to heated argument. One thing is clear, however, that conversion of taxon versus character data into distance measures involves discarding an enormous amount of information and should be avoided when possible.

Methods for tree construction from distance data can be broadly divided into three categories, viz. hierarchic clustering based on overall similarity (i.e. phenetic methods such as **UPGMA** and **WPGMA** (respectively, unweighted and weighted pair-group methods using arithmetic averages; see chapter 4), the **distance Wagner** procedure (Wagner, 1961; Farris, 1972; Swofford, 1981) and its variations, and the **Fitch-Margoliash** method (Fitch and Margoliash, 1967). However, within these categories many possible methods have been proposed (Swofford and Olsen, 1990). The efficacies of these and several other approaches depend to a great extent on whether or not the distance data themselves are additive.

Of the above methods for coping with distance data, the latter two procedures are based on underlying cladistic principles in that they attempt to minimize the amount of change that must have taken place to produce the available distances. The Fitch–Margoliash method is based on trying to find the shortest possible tree that can span the set of taxa given their relative distances. Initially it uses a phenetic clustering procedure (such as UPGMA; see section 4.3.3) to obtain a first estimate of the tree, but then tries to improve on this by measuring many rearrangements based on the first tree. The more rearrangements tried, the more likely it is that the best tree obtained will be the shortest one (Farris, 1981), but of course, there is no guarantee that it is unless all possibilities have been tried. The distance Wagner method differs from the Fitch–Margoliash method in that the tree construction process itself involves the sequential addition of taxa such that at each stage in its construction the partial tree has minimum length.

When more than one method is available for analysing a particular type of data, it seems reasonable to ask which is better. However, as pointed out by Swofford (1981), the answer may depend on the criteria used to define the word better. This is very much the case with methods for analysing distance data.

3.2.4 *Rooting trees*

Rooting a tree is the process whereby a decision is made as to which of the tree's nodes is evolutionarily closest to the common ancestor of the taxa concerned and thus represents most closely the taxon from which all other members of the group are derived. In practice, the rooting procedure either directly or indirectly involves assigning polarity to at least some of the character states displayed by members of the group. Direct methods involve consideration of the likely polarities of individual characters based on various reasonings (see section 2.2.2), whereas indirect methods make use in the analysis of one or more additional taxa which are believed to form a sister group of the group under investigation. The distinction between these approaches is not sharply defined. It should also be noted that while most computer algorithms for finding parsimonious trees initially produce unrooted ones which need subsequent rooting for interpretation or output purposes, rooted trees have to be calculated when characters in the analysis are not equally likely to change from one state to another as to change in the reverse direction.

Having decided on an outgroup (or set of outgroups) or on the character states of an hypothetical ancestor constructed through consideration of one or more outgroups (or other methods; see section 2.2.2), parsimony analysis can be applied either just to the set of taxa in which we are interested or to these plus the outgroup(s)/ hypothetical ancestor. Whether or not to include the ancestor within the phylogenetic analysis itself has been debated. Inclusion of an ancestor can sometimes lead to shorter trees than when an ancestor group is subsequently tacked on to an already minimal length tree constructed without the inclusion of an ancestor. Further, since real evolution necessarily must entail an ancestor it seems reasonable that an ancestor should also be included in the analysis.

For analyses performed without an included ancestor, rooting is usually done by finding a node where likely character states would most closely resemble those of a hypothetical ancestor and then inserting the root there. In practice, this amounts to finding a position where the ancestor can be attached that increases the tree length by the least amount. This process, usually referred to as **Lundberg rooting**, unfortunately often yields several equally acceptable (i.e. equally ancestor-like) nodes and so still leaves the problem of choosing between these.

3.2.5   *Consistency and other indices*

Having obtained a tree or trees through analysis of a particular data set, it is relevant to enquire how well the distribution of character states is explained by the tree. Several indices have been proposed to describe the closeness of fit of a data set to the trees obtained from it. In effect, these indices express the degree to which the observed variation requires homoplasy. In general, indices may be applied either to a single character, or to all of the applicable characters considered together thus giving an overall perspective of the fit of the data to the tree. Such indices based on the whole set should normally be referred to as **ensemble indices** to distinguish them from single character indices, but since ensemble indices are the most widely used, the 'ensemble' is frequently dropped. In the following, individual character indices are denoted by lower case identifiers while ensemble indices are denoted by capitals. Three indices in widespread use are described below together with some notes on their particular properties.

The first and probably the most frequently employed goodness of fit index is the **consistency index** (c) which was defined by Kluge and Farris (1969) as

$$c = m/s$$

where m denotes the minimum number of character state transitions that a character could show on any tree, and s is the actual number of transitions required on the tree in question. Calculation of c is illustrated in Figures 2.5–2.9. For example, in Figure 2.6 the right-hand tree requires four character state changes but it is possible to arrange the character states such that only two changes are needed as shown in the left-hand tree. Thus the consistency index for the character in the right-hand tree is 2 divided by 4, i.e. $c = 0.5$ while for the other tree $c = 2/2 = 1.0$.

If the tree requires no homoplasy then $c = 1$, lower values of c indicate progressively poorer fits.

The ensemble consistency index, C, defined as

$$C = \Sigma\, m/\Sigma\, s$$

gives a reasonable impression of the data. However, it is influenced by the number of **uninformative** characters present and so this can be misleading if it is necessary to compare indices obtained either from different suites of characters or when different weighting systems have been used. Uninformative characters should therefore be excluded from the calculation of C, and if this is done, then it provides a good indicator of fit provided that it is used comparatively and for unweighted data (Carpenter, 1988). In practice ensemble consistency indices are frequently found to be far lower than might be hoped for, values between 0.25 and 0.35 being common and very high values (0.8–1.0) may indicate that the authors have rejected certain homoplastic characters before carrying out their analyses.

In an attempt to overcome the constraints on C, Farris (1989) proposed two other indices, the **retention index** (r) and the **rescaled consistency index** (rc; see below). The single character retention index is defined as

$$r = (g-s)/(g-m)$$

where g is the greatest number of character state transitions that a character could show on a given tree. The ensemble retention index, R, being given by

$$R = \Sigma \ (g-s)/(g-m)$$

The retention index can be thought of as the proportion of the apparent synapomorphy that can be accepted as true synapomorphy, i.e. excluding that part which can be attributed to homoplasy. Thus if $s = m$ then $r = 1$ and there is no homoplasy; however, if $g = s$ then $r = 0$ and all apparent synapomorphy can be considered as being due to homoplasy. Uninformative characters therefore give $r = 0$ and thus these do not influence the index obtained by considering all the available characters (both informative and uninformative) together.

The rescaled consistency index is defined as the product of the retention index and the consistency index (i.e. RC), and has the desirable property that it makes allowance for the effect of differences in the value of g for characters that otherwise show similar levels of homoplasy.

In general, the more characters or more taxa that are included in an analysis, the more homoplasy will be encountered because few characters are perfect indicators of phylogeny and because if a worker only uses a few characters then the ones selected will generally be the least homoplaseous (or at least the most congruent) ones available (Archie, 1989a,b). This is illustrated in Figure 3.5 which shows the general distribution of C versus the number of characters together with separate regressions for molecular and morphological data (Sanderson and Donoghue, 1989). However, it is reasonable to assume that the distribution of even a rather homoplastic character might provide some phylogenetic information, and therefore such characters should not

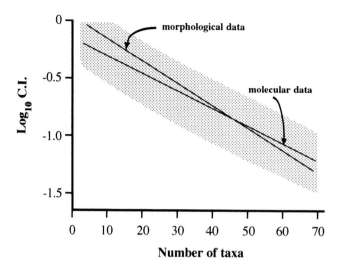

Figure 3.5 An illustration of the general negative relationship found between consistency index (c.i.) here represented by its logarithm, and the number of taxa employed in the study (simplified after Sanderson and Donoghue, 1989). Original results were based on an analysis of 60 data sets, and trends for molecular and morphological data sets are also illustrated independently.

necessarily be excluded from consideration although, of course, they would be excluded in compatibility analyses. To prove the point Archie (1989a) examined the effect of randomising character state assignments on consistency indices for a range of real data sets. His results show that characters giving low consistency indices are nevertheless still informative, i.e. randomisation leads to even lower consistencies. However, a slightly different view is postulated by Faith and Cranston (1991) who found several instances in which data appeared to have such a large random element that little confidence could be placed in any of the trees obtained.

Unfortunately, the negative correlation of consistency index versus data set size shows it to be a poor measure for comparing homoplasy between data sets. In an attempt to overcome this problem, Archie (1989b) proposed a different sort of measure which he termed the **homoplasy excess ratio** or **HER**. Calculation of the HER in its pure form requires knowledge of the observed homoplasy excess (HE $= S - M$) and a value which he termed the maximum homoplasy excess (MHE).

The MHE was defined as the homoplasy excess (i.e. the number of homoplastic changes) when character states of each character in a given dataset were randomized across the taxa and this randomized data used to generate new minimum length trees. Thus the MHE approximates the homoplasy excess if the data contained no phylogenetic information. The HER was thus defined as:

$$HER = 1.0 - \frac{HE}{MHE}$$

However, randomizing character states is messy and time-consuming and based on an empirical analysis of many real datasets; Archie was able to provide some short-cut methods for evaluating the HER.

3.2.6   *Weighting characters*

Although numerical taxonomic methods were originally intended to overcome the potentially biased views of taxonomists in favour of particular characters by assigning all informative characters equal importance, it is clear that there are great differences in phylogenetic significance between different character types. For example, changes at the third position of a codon are common and will frequently be selectively neutral whereas particular chromosomal inversions may be less frequent and highly unlikely to undergo reversion. Such differences in the information content of characters, especially when they are so large, clearly ought to be reflected in phylogenetic analysis by assigning different weights to different characters. The problem is therefore not so much whether weighting is justified but how to decide on what actual weights could justifiably be applied. As pointed out by Fitch (1984) even the process of selecting characters for inclusion in an analysis is a process of weighting wherein those characters that are excluded are effectively being given a weight of zero, leading to the negative empirical relationship observed between C and the number of characters used (see Figure 3.5).

Unfortunately, it is frequently difficult or impossible to determine what numerical weighting value would reflect the true difference in the informative value of a pair of characters, and therefore many taxonomists have opted not to assign weights *a priori* and instead to rely on analysing the largest possible set of characters without weights.

The use of as many characters as possible stems from the idea that even quite homoplaseous characters do still in fact carry phylogenetic information albeit confused with additional transitions. That this may be so was shown by Archie (1989a) (see above). However, a consequence of this is that inclusion of more and more characters with their almost inevitable addition to homoplasy will lead to an increase in the number of equally parsimonious trees obtained.

Another aspect of weighting is less deliberate and stems from the fact that some characters used in phylogenetic analysis are binary while others may be multistate. Parsimony analysis will consequently attribute more weight to multistate characters than to binary ones because multistate ones will contribute more to the length of the tree and hence will have a stronger influence on the final most parsimonious one. This is a philosophically difficult issue to resolve as it begs the question as to whether we should treat characters or character states as having equal weight in parsimony analysis. This author supports Farris (1990) in that any *discrete* character state change should be weighted equally if for no other reason than there does not seem to be any good reason why a character that has by chance undergone several potentially informative evolutionary trans- formations should have these transitions down-weighted. However, when it comes to discretely coded continuous characters (Baum, 1988; chapter 2) the number of transitions is dependent on the coding system employed and therefore it may be desirable to standardize these so that each character rather than character state is given equal weight. It should be emphasized that the issue of weighting is by no means resolved (see Colless, 1980).

A little-studied aspect of weighting has its origins in compatibility analysis. The argument of compatibility analysis is that characters that do not truly reflect evolution in a group of taxa are likely to be in conflict with characters that do. Thus if most characters are true indicators of phylogeny then the number of other characters that are compatible with a given character may be related to how truly it represents evolutionary history, i.e. the more homoplasy a character displays then the fewer other characters with which it is likely to be compatible. The main criticism of compatibility analysis is that in discarding characters that perhaps show only a small number of homoplaseous changes, much information is being discarded. Com- patibility can, however, be used to generate a differential weighting system that will retain some otherwise discarded data, and Sharkey

(1989) has suggested a possible method for weighting characters based on their compatibility. This weighting system is independent of the final tree since the character compatibilities are derived directly from the raw data. In contrast, Carpenter (1988) has proposed a system for weighting characters according to their fit on an initially calculated set of parsimonious trees. This *a posteriori* method which is useful for reducing the set of equally parsimonious trees and increasing their resolution is described in the following section.

3.2.7   *Coping with multiple trees*

Although some data sets will yield a single most parsimonious tree, it is not uncommon for several to many hundred or thousand different equally parsimonious trees to result from the analysis of a given data set. All of these will usually have some features in common, i.e. arrangements of particular subsets of taxa that are supported by all trees, but the relationships of other taxa may be variable, and the overall appearance of the trees may vary greatly. As a corollary to this, inclusion or exclusion of a single character or a single taxon from an analysis can potentially have a profound effect on the appearance of the most parsimonious tree (Fitch, 1984).

There are essentially two strategies for coping with these multiple solutions. Ways may be found to justify choosing one of the equally parsimonious trees from the remainder as that most likely to represent the true phylogeny, or the user may accept that some parts of the classification cannot be resolved satisfactorily at present and therefore may concentrate only on those aspects of the trees that are consistently revealed. The former strategy usually involves character weighting arguments whereas the latter involves **consensus** techniques (see section 3.2.8).

It has been argued that any of a set of equally parsimonious trees is better than a consensus tree because the greater resolution of the latter means that it has less explanatory power than any more fully resolved tree (Carpenter, 1988). Many workers to date have preferred the conservative consensus approach if for no other reason than that it gives a feeling of security and confidence about the information it presents (e.g. Hillis, 1987). However, Carpenter presented a successive approximations weighting approach that provides a way of using the original data to help narrow down the choice of trees, sometimes until only one is left. His method is based on the fact that multiple trees result from the confusing effects of homoplaseous characters. Therefore, if there is a way of deciding what characters are showing most homoplasy, these could be afforded less weight than the other characters and so both reduce the effects of their homoplasy and the number of equally parsimonious trees obtained. In Carpenter's iterative method firstly unweighted characters are used to obtain a set of most parsimonious trees and then the consistency index of each character is averaged over all trees and the resulting value used to provide a weighting for that character which is then used in a subsequent parsimony analysis. The process is then repeated until the resulting set of trees from successive runs remains unchanged. Several variations on this general paradigm have been presented; for example use of the retention index or rescaled consistency index instead of the consistency index or use of the best consistency value of each character rather than the mean, etc. At present too little is known to decide which system is better.

However, successive approximations character weighting, while elegant, does have some drawbacks. First, with incomplete data sets (or ones in which there is a lot of intrataxon polymorphism) some consistency index calculations might give unduly high values and therefore bias the outcome of the tree selection process. A second consideration was pointed out by Sharkey (1989), namely that the successive approximations procedure is inherently an *a posteriori* **weighting** method, that is, the weighting system adopted depends on the outcome of the first parsimony analysis. As an alternative, Sharkey proposed that weighting, if adopted, should not depend on any particular tree but instead be a function of the unanalysed data set, in other words it should be an *a priori* **weighting** method. In *a priori* weighting, character compatibility calculated at the beginning is used to give more weight to characters which are compatible with more other characters. Implementation of Sharkey's system is unfortunately not easy and cannot be implemented directly on any current parsimony package.

3.2.8 *Consensus trees*

Once a taxonomist has obtained a number of trees for a given data set, an obvious question that springs to mind is what features if any do these trees have in common and conversely what particular relationships differ from tree to tree and therefore may need further investigation. The commonest way of investigating this is to construct a so-called **consensus tree** which depicts those relationships that were constant in the set of original trees, and thus gives the worker an idea as to what aspects of the results are best supported by the data.

Several different consensus techniques are commonly employed and, depending on the variation present among the set of trees from which the consensus tree will be calculated, these may all yield the same tree or each may produce a different consensus tree. As Smith and Phipps (1984) explain, consensus techniques fall into two categories, cluster counting methods and cluster intersection methods. The first of these involves counting the clusters of taxa that are supported (or sometimes not contradicted) by the initial set of trees. The second group of methods are based on set theory and identify consensus clusters on the basis of set intersections. Of the former type, the three most commonly encountered methods are **strict**, **semi-strict** and **majority rule**. The most commonly used of the second group is the **Adams** consensus method although Smith and Phipps also briefly describe another called the **durchschnitt** consensus method.

Examples of the results of applying four of the most commonly used consensus techniques to a set of three initial trees are shown in Figure 3.6 and these are described briefly below. Most techniques are equally applicable to rooted and unrooted trees but the algorithm proposed by Adams (1972) can only be applied to rooted ones. It should be noted that cluster counting methods such as the first three only show as consensus groups those groups that are present in at least half of the initial trees while cluster intersection techniques can show sets of taxa that never appear as groups in the starting trees.

Strict consensus trees show as resolved only those features that are supported by all of the initial trees, thus in Figure 3.6 only the grouping of taxa e, f and g is distinguished from the remaining taxa as they occurred as a monophyletic clade in all three of the initial trees. Relationships within this group were not resolved in the consensus tree because they were not resolved in initial tree number (ii) even though both other initial trees had e as the sister group of $(f + g)$. Taxa a, b, c and d in the initial trees show no consistent set of relationships and thus originate together from a polychotomy in the consensus tree together with $(e + f + g)$.

The semi-strict consensus tree differs from the strict tree in that it shows those features that are resolved in all the initial trees or are resolved in some of the initial trees and not contradicted in the others.

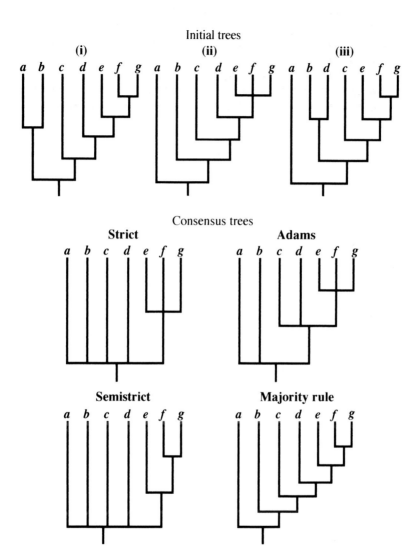

Figure 3.6 Four different types of consensus trees (*below*) constructed from the three different initial trees (i) to (iii) (*above*).

Thus in the example shown in Figure 3.6, taxon e is now seen to be the sister group of $(f + g)$ because it is supported in initial trees (i) and (iii) and is not contradicted by the polychotomy in tree (ii).

Majority rule trees show those features that are supported by an arbitrary majority of the initial trees even though these features may be wildly contradicted by the relationships suggested in a minority of the initial set. Thus in the example shown, all taxa are resolved with respect to one another in sequence even though no one of the initial trees shows this arrangement. Any majority cut-off can be arbitrarily chosen, thus the consensus could for example, be based on those features present in 80% or 90% of trees, though it is difficult to justify any majority rule approach since, on grounds of parsimony, all trees are equally short and therefore equally improbable.

Construction of Adams consensus trees is based on the idea that the common ancestor of a group of taxa in a consensus tree should be set at the furthest distance from the origin at which it occurs in all the initial trees. Thus in the example provided, taxa c and d have a common ancestor one step from the base (taxa a and b) because in none of the initial trees does either diverge from the stem closer to the origin than either a or b.

It is clear that in the example chosen, different consensus techniques have yielded markedly different arrangements. Which is best is a far more difficult question to answer. While some would argue against any consensus tree due to its lesser explanatory power (Carpenter, 1988), the appeal that a consensus tree summarises some invariant aspect of the information content of the original set of trees will no doubt remain with us for a long time.

3.2.9   Comparing trees

While the similarities between two trees or two classifications can be roughly assessed by eye, it may be desirable to obtain a more objective interpretation. For example, the congruence or lack thereof of two or more trees obtained from different data sets may provide information on the relative values of different types of data and on the likelihood that the trees represent the true ancestry of the group.

Early attempts at comparing dendrograms such as that by Rohlf (1963) made direct use of the taxa versus taxa matrices of similarity coefficients used in phenetic analysis rather than by comparing the trees themselves. However, the similarities between pairs of trees can be quantified directly, and with less loss of information, by considering the sets of taxa that are present beyond any point in either of the two trees being compared, irrespective of any differences in the relationships between these taxa in the given trees. For a pair of randomly constructed trees there will be a only a low probability of the same set of taxa occurring beyond any given point, whereas for two very similar trees there will be many points beyond which the sets of taxa will be identical. As was shown in section 3.2.2 there are $2t - 3$ links in an unrooted fully bifurcating tree with t terminal taxa, therefore there are $t - 3$ internal links, cutting of which will lead to two sets of taxa each with two or more elements. If the number of sets of taxa that the pair of partitioned trees have in common is denoted by m, then m must range from 0 to $t - 3$. Thus by partitioning one tree into two and then examining whether either of the two sets of taxa thus formed can also be obtained by any binary partition of the other tree, it is possible to quantify what proportion of the total possible number of splits yield the same sets and therefore derive a measure of overall tree similarity (Penny et al., 1982, 1992). Some other measures of congruence are discussed by Sackin (1985).

3.3   Which method? — an overview

As outlined above there are several competing methods employed for cladistic analysis and while at present parsimony methods seem the most favoured for many sorts of data there is no guarantee that other methods might not be developed in the future or that parsimony methods are necessarily best in all circumstances. If a single method were always capable of giving the correct result, in this case revealing the true phylogeny of the group of organisms concerned, then for any group of organisms that method would always yield the same trees, no matter what set of characters were employed. While unrealistic, this principle does have a useful corollary which is that poor techniques, i.e. ones that seldom yield the correct phylogeny, will probably tend to produce a range of different cladograms.

An interesting and cautionary story is discussed by Fitch (1984). This concerns an analysis by Baum of a number of cultivars of the cereal crop oats, whose ancestry over the last 100 years was completely known. The data set obtained from these was then used to test the effectiveness of a variety of tree-constructing procedures with the result that no procedure yielded a correct reconstruction. A corollary of this was that none of the characters employed was found to be 100% consistent, i.e. all displayed some homoplasy.

In the above survey of parsimony methods one prime assumption has been that character state transitions are intrinsically unlikely, and therefore homoplasy is particularly unlikely. Unfortunately, in the real world, homoplasy is often commonplace. For example, a brief consideration of flower colours or leaf shapes will show that for practical purposes the same or indistinguishable colours and forms recur throughout the plant kingdom. Because of this, taxonomists would not normally include either of these in an analysis of higher level plant relationships. In other words, they would apply a filter on the characters selected for analysis, rejecting those considered to have low information content. This is no less than an extreme case of *a priori* character weighting.

To quote Felsenstein (1982a), 'It is essential that we do not adopt a single method as a universal panacea, but that an attempt is made to understand the biological assumptions and statistical behaviour of each method.'

3.3.1   *How well does parsimony analysis estimate trees?*

In trying to determine how good current phylogenetic methods are at determining true phylogenetic relationships we enter the realm of the circular argument in that we would need to **know** a true phylogeny in the first place. One way around this is to use simulation to create phylogenies through the implementation of a realistic evolutionary model, and then on the basis of the character states of the final field of taxa to attempt to back construct their phylogenies. However, even this is not as simple as it may seem because we do not know exactly how evolution and character transitions take place. In an attempt to allow for this Rohlf *et al.* (1990) used simulations to model the evolution of a monophyletic group of eight taxa under three separate assumptions of how characters change with splitting (e.g. speciation) events: **phyletic** — character states can change at any time; **speciational** — characters change only at splitting events; **punctuational** — characters change only at splitting events and then only in one of the two daughter lineages. In agreement with Fiala and Sokal (1985), maximum parsimony (cladistic) methods out-performed phenetic clustering but even the minimum length trees were frequently 'wrong' in that they did not show the evolutionary pathways of the model. Two factors were found to influence the accuracy of the phylogenetic results, the evolutionary model itself and perhaps surprisingly, the topology of the evolutionary tree. In particular, models based on the idea of punctuated equilibrium evolution (Eldredge and Gould, 1972) were found to be particularly difficult to reconstruct accurately. Rohlf *et al.* concluded from their results that the great majority of proposed phylogenies are likely to be inaccurate and suggested that it would be more sensible to refer to all proposed phylogenies as estimates rather than, as is often the case, actual reconstructions.

3.3.2 *Compatibility versus parsimony*

Compatibility has been attacked by many workers because it usually results in much potentially informative data being discarded. However, in some ways compatibility agrees better with Hennig's original cladistic methodology in that his method overcame homoplasy by identifying characters that were in conflict and then re-examining these until the *error* was detected and so could be corrected. Thus Hennig's methodology was essentially one in which only compatible characters were employed. Perhaps more relevant to the debate over whether compatibility is as justified as parsimony is the realization that under certain circumstances of character weighting, parsimony analysis actually becomes identical to compatibility analysis (Farris, 1969; Felsenstein, 1983).

A reverse side of considering compatibility analysis comes from the common observation that the largest clique often comprises depressingly few characters (Fitch, 1984). This is sometimes taken by proponents of compatibility analysis to show that the characters are so homoplaseous that it would be foolish to try to draw too many (if any) taxonomic conclusions from them. The counter argument of those who favour parsimony analysis is that such data nevertheless do carry considerable phylogenetic information, albeit noisy information, and that their method enables this to be put to best use. Whichever argument you may side with, it is likely that the idea that we can reconstruct any phylogeny with absolute certainty is at present untenable and that it is more realistic to consider cladograms as estimates of history, in which case the question then posed is how can we judge how likely our estimates are to be the truth?

3.3.3   Congruence between data sets (or how do we know when to believe a phylogeny?)

Since any group of organisms can have had only one evolutionary history it follows that any combinations of analytical method and data set that yield the true phylogeny will necessarily agree. Conversely, if two phylogenetic hypotheses disagree then at least one must be wrong. Agreement between trees is usually termed **congruence** and obtaining congruent trees from more and more independent data sets for a particular group of organisms gives stronger and stronger support for the evolutionary hypothesis to which they lead. Unfortunately, obtaining congruent results even for small numbers of taxa belonging to well known groups with good fossil records can be surprisingly hard.

Independent data sets include, for example, larval versus adult characters (Michener, 1977), gametophytic versus sporophytic (Rohrer, 1988), animal artefacts versus morphology (Wenzel, 1991), chromosomal versus morphological, molecular versus morphological characters (DeSalle and Grimaldi, 1991) and many other possible combinations (Sytsma, 1990; Smith, 1992). In the case of larval and adult characters, for example, conflicts (lack of congruence) might be expected to result from quite different selection pressures acting on larval and adult characters giving rise to environment-dependent parallelism or convergence though this does not mean that larval and adult classifications will always be in conflict (Rohlf, 1963). On the other hand, as most molecular sequence data is generally believed to be less subject to convergence or parallelism at the individual residue level than morphological characters, conflicts between morphological and sequence data tend to result in pressure on the morphotaxonomist to reappraise their data set identifying likely sources of error *as necessary* (DeSalle and Grimaldi, 1991). Errors in tree construction can, of course, also result from sequence data itself if it has been inappropriately analysed, these are often referred to as apparent discrepancies.

Smith (1992) discusses some problems connected with working out the evolutionary relationships between the five extant classes of echinoderms (sea urchins, starfish, sea cucumbers, etc). Initial studies on molecular data sets produced some surprising results compared with what seemed to be well founded phylogenies based on morphological and palaeontological evidence. Here the lack of congruence was shown to result from a combination of the use of distance measures rather than of individual nucleotide position data, and on the choice of an outgroup whose sequence had diverged so much from the echinoderm sequences under study that it was effectively random. Other problems have been determined as resulting from the choice of DNA sequences that were inappropriate for assessing higher level relationships because either too many homoplaseous changes had accumulated over the time periods involved or too few changes had accumulated overall.

3.3.4 *Reticulate evolution, hybrids and intraspecific evolution*

One of the basic tenets of all current phylogenetic analysis methods is that the underlying evolutionary processes that they are endeavouring to unravel are essentially bifurcating or at the very least, are not **reticulate**. In other words, species are expected to become split during evolution resulting in pairs of daughter species but they are not generally expected to arise from hybridization events between two separate parent species. Similarly, **introgression**, where genes from one species become transferred to another via a hybridization event followed by subsequent backcrossing, is believed to be insignificant and likewise the possibility of horizontal gene transfer is usually ignored (Smith *et al.*, 1992). Unfortunately, even now not much is known either about the frequency with which introgression occurs or how often hybridization events give rise to new species in the wild. Certainly, in captivity or under cultivation some groups of organisms appear to be fairly happy to hybridize and, usually as a result of an associated polyploidy event, the resulting hybrids may be quite capable of reproducing. Indeed, many cultivated plants including crops have undoubted hybrid origins (see also section 7.6). However, it is generally agreed that these phenomena are more common among higher plants than among animals (Orr, 1990; Heron, 1992; Thompson and Lumaret, 1992) but also see Mayr (1992).

When data pertain to different populations within a species phylogenetic analysis may be frustrated by the occurrence of gene flow between these populations. With separate biological species this is (by definition) not a problem and it can be assumed that gene complexes will have been evolving effectively independently in each lineage after their initial speciation event. However, within a species this cannot be assumed and it is possible that gene flow between populations could result in the transfer of whole groups of genes from one population to another otherwise only distantly related one. A prime example where such effects could have a profound influence can be found in the several recent attempts to unravel the history of the various human races using molecular genetic techniques (Cann *et al.*, 1987; Gibbons, 1992).

3.4   Cladistics and classification

Probably one of the most contentious aspects of cladistics does not concern its value in detecting or estimating the true phylogeny of a group but rather on how far the phylogenetic hypotheses that result should be applied to formal classification and to nomenclature. This problem comprises three rather separate issues. First, there is the fundamental question of whether it is desirable to follow phylogeny rigidly in setting up a classificatory system. Second, if a phylogenetic basis is accepted then what criteria will be employed in deciding on where to draw the line between higher taxa and what ranks to afford them. Third, whether the phylogenetic hypothesis being used will ultimately be stable as replacement by a revised (improved) hypothesis could lead to nomenclatural instability (Erzinclioglu and Unwin, 1986; Claridge, 1991).

To illustrate the first of these issues, it is worth considering a widely appreciated example of the possible confusion that could result from strict adherence to cladistic rules in naming higher taxa. To the layman and to many biologists, the class Reptilia includes a small number of animal groups, the lizards and snakes (order Squamata), the turtles, tortoises, terrapins, etc. (order Chelonia), and the crocodiles, alligators, caymans, etc. (order Crocodilia). To these may also be added the extant, lizard-like tuatara from New Zealand (order Rhynchocephalia) and a number of extinct groups. These animals have a general feel about them, they share many features such as a hairless skin, they are cold-blooded, they lay eggs with a lot of yolk (with a few exceptions), their embryos develop within an amnionic and chorionic membrane, etc. In contrast, the birds (class Aves) and the mammals (class Mammalia) appear very distinct from the reptiles; the former have feathers, the latter fur and they are both warm-blooded (homeothermic) (Kemp, 1988).

However, cladistic analyses and fossil evidence do not support the idea of the Reptilia forming a monophyletic clade separate from the birds. Instead, it is fairly clear that the Crocodilia are evolutionarily closer to the birds than they are to the other reptiles, in more precise terms, Crocodilia + Aves forms a monophyletic clade separate from the other reptiles. Thus, if cladistic principles are brought into play, the class Reptilia would have to be redefined and a new name created to include the birds, crocodiles and probably dinosaurs.

To many people, formalizing such phylogenetic findings by creating new names and eliminated well-known and seemingly biologically informative names such as Reptilia, no matter how securely based, serves little purpose. They argue that it adds unnecessarily to the already vast array of names and that in 'sinking' names such as Reptilia, a potentially informative term is removed. It may be well to remember that whereas polyphyletic groups were never monophyletic (see chapter 2), paraphyletic ones were until such time as one of their included species gave rise to such a distinct (distinctly different) group that it was deemed to warrant its own higher taxonomic status (thus the class Mammalia evolving from the Reptilia). In my opinion therefore, paraphyletic taxa are acceptable so long as people recognize them as such, although many other cladists would disagree. Polyphyletic taxa on the other hand are far harder to defend.

Another important feature of a classification is that it should be testable. This can only be the case if the classification is based on the true **phylogeny**, or in other words evolutionary history, of the group. If a classification truly reflects phylogeny, then it should be supported by all (or at least most) new characters (data). If newly discovered characters do not support the existing classification, then the latter may need to be re-evaluated. Unfortunately, the need for re-evaluation may be more common than many would care to accept.

The second major problem with applying cladistic principles rigidly to classifications is a purely practical one, namely that if names in taxonomy and systematics are to be applied to monophyletic groups, where are we to draw the line between taxa to be named. Clearly it would be ridiculous to apply a distinct name to every monophyletic clade within a phylogeny because the number of names required would be astronomical. From the opposite point of view, pairs of monophyletic sister groups can be continuously united creating larger and larger units thus posing the problem of how big a group should be or what other criteria should be employed for it to receive a subfamily, family, superfamily name, etc. Suppose that two monophyletic groups previously treated as, say, separate subfamilies are shown conclusively to be sister groups. Should they now be merged into a single subfamily or not? In practice, the decision will depend on a number of factors, the size of the two groups, the morphological gap between them and the amount of information that keeping them separate may carry about, for example, their biologies. Hennig proposed that the taxonomic category of a group should be determined by its age, but given the problem of determining age for many groups (see chapter 12), such a system would hardly be expected to stabilise for a long time. Further, different taxonomists often have different proclivities to combine or separate taxa, so-called **lumpers** and **splitters**. In the end such decisions are at present effectively arbitrary and are likely to remain so for some while although Sorensen (1990) offers some practical suggestions for partitioning phylogenies.

The third problem with naming all higher taxa on the basis of pure cladistic principles is that for most taxa we do not yet have a definitive phylogenetic reconstruction. If cladistic methods were nearly always capable of yielding the true phylogeny of a group this would not be much of a problem; after all, there is only one true phylogeny and therefore if this were known there would be only one cladogram on which to base the classification. Unfortunately, taxonomists are not infallible and not infrequently make mistakes of various types, especially in character interpretation (Robinson, 1986), and, furthermore, even if all aspects of an analysis are performed perfectly and there is a sufficiency of characters, the resulting cladogram is probably unlikely to be totally accurate, although with luck there should be substantial areas in which the cladogram is true.

CHAPTER FOUR

PHENETIC METHODS IN TAXONOMY

4.1 Introduction

As pointed out in chapter 1 (Figure 1.1), phenetic clustering methods, in which taxa are grouped on the basis of overall similarity, neither provide reliable evidence of evolutionary relationships, nor form a sound basis for classification. Nevertheless, they still have several important roles in taxonomic work, particularly in identification and in determining whether samples of specimens comprise one or more distinct entities. Because pheneticists and their methods frequently deal with individuals or populations as well as with species and higher taxa, they often refer to **operational taxonomic units (OTUs)** as a convenient cover-all term.

Most phenetic methods of concern to us here involve taxon to taxon **distance, similarity** or **dissimilarity** measures rather than raw taxon × character matrices. As their name implies, distance and dissimilarity measures increase with the dissimilarity between taxa while similarity measures decrease with dissimilarity. Distance and dissimilarity are sometimes treated as the same thing though a distinction between them can be made. Thus, since similarity is usually measured on a scale of 0 to 1 with 1 equal to perfect identity, dissimilarity is often defined as 1-similarity. However, while two OTUs that are identical are separated by a distance of 0, there is no standard upper limit to distance ranges. Swofford and Olsen (1990) suggest three transformations for converting a similarity, s, (measured on a 0 to 1 scale) to distance, d, thus:

$$d = 1 - s \text{ (i.e. the same as dissimilarity) [range 0 - 1]}$$
$$d = -\ln(s) \qquad \text{[range 0 - \infty]}$$
$$d = (1/s) - 1 \qquad \text{[range 0 - \infty]}$$

4.1.1 *Similarity and distance measures*

Two organisms or samples of organisms will usually differ to a greater or lesser extent and for many reasons it may be desirable to quantify the degree by which they differ from one another or, conversely, how similar they are. Such measures are essential for many automated identification systems such as are often used to identify pathogenic organisms. Probably because much early work on these measurements was concerned with discrete morphological characters and all or nothing physiological properties of bacterial cultures such as their abilities to ferment various sugars, many of the numerous similarity and distance indices proposed only use binary characters though a few can accept both binary and continuous measurements. In fact, so many different similarity or distance measures have been proposed that it would be of little value to list them all here, instead the following small selection includes some of the most widely used. For a fuller coverage, the reader can consult Clifford and Stephenson (1975) or Austin and Priest (1986).

4.1.2 *Measures using binary characters*

Binary characters can be dealt with in two ways depending on whether they can be obviously polarized, for example in terms of the presence or absence of a fermenting ability, or whether the character simply has two states that it is undesirable or impossible to classify in terms of presence or absence. Several indices proposed for bacteriology need data of the former type because they accord extra weight to positive than to negative results. Data are presented in the form of a truth table of test results between the two taxa or operational taxonomic units (OTUs) thus

$$
\begin{array}{c|cc}
 & \multicolumn{2}{c}{\text{OTU 1}} \\
 & + & - \\
\hline
+ & a & b \\
- & c & d \\
\end{array}
$$

OTU 2

and the indices make use of the number of tests giving the results a, b, c and d.

The **Simple matching index** (S) or **coefficient** is defined simply as the number of tests giving identical results (a or d) divided by the total number of tests, thus

$$S = \frac{a + d}{a + b + c + d} \quad \text{(range 0 to 1)}$$

This index makes no distinction between +/+ matches and −/− matches. Several modifications on this theme which also take shared negative scores (−/−) into account have been defined in order to give differential weighting either to +'s or matches. Two examples will suffice, Russel and Rao's, S_R, and Sokal and Sneath's, S_{SS}, which are defined as

$$S_R = \frac{a}{a + b + c + d} \qquad S_{SS} = \frac{2(a + d)}{2(a + d) + b + c}$$

In contrast, some other similarity indices ignore −/− matches, for example the Jaccard index and the Czekanowski (or Dice, or Sorensen) index, S_J and S_C, respectively, which are defined as follows

$$S_J = \frac{a}{a + b + c} \text{and } S_C = \frac{2a}{2a + b + c}$$

Likewise, the simplest distance measure usually called the **total difference** simply takes the number of tests giving different results in the two samples divided by total number of tests, viz.

$$\frac{b + c}{a + b + c + d}$$

A slightly more complex measure is the Itamann index, S_I which emphasizes the importance of similarities over differences by subtracting the total difference from the simple matching index, i.e.

$$S_I = \frac{(a + d) - (b + c)}{a + b + c + d}$$

4.1.3 *Distance and similarity measures using continuous data*

Taking into account continuous variables is complicated because differences in the ranges of values obtained from different variables means that each would contribute unequally to simple indices such as those listed above. Overcoming this requires some sort of adjustment procedure and in turn these generally require knowledge of character variation in a large set of OTUs. The most commonly employed method for equalizing the effects of different continuous variables is called **normalization** and involves two stages. First the mean and standard deviation is calculated for the values of each character (all taxa combined) and then each value for a particular character has the respective character mean subtracted and the result is divided by the standard deviation. The resulting values for each character have a mean of 0 and a standard deviation of 1.0.

The commonest distance measure for a pair of taxa based on a set of normalized continuous variables is the **Euclidian distance**. This is equivalent to the real physical distance by which the taxa would be separated if they were plotted in a space in which each axis represents one normalized variable and each axis is perpendicular to all others. Calculation of Euclidian distance is illustrated in Figure 4.1. In the left-hand illustration only two characters are involved and the distance between the two taxa is simply given by Pythagoras' theorem. Exactly the same process can then be used to calculate the Euclidian distance between two taxa for three or more variables as is shown in the right-hand figure.

Unfortunately, most distance measures are restricted either to using only discrete data or to using only continuously variable data. The generalized coefficient of similarity described by Gower (1971) can handle data sets comprising both types (Sneath and Sokal, 1973).

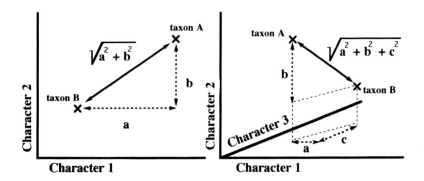

Figure 4.1 Illustration of the calculation of the shortest distance between two points in two dimensional (*left*) and three dimensional (*right*) space. The idea can be extended to any number of dimensions and the shortest distance is referred to as the Euclidean distance.

4.2 Analysing similarity and distance data

Essentially there are two main uses for distance data in numerical taxonomy. First, it can be used to search for patterns in the data so as to gain an insight into what groups may be present among a given set of OTUs and how similar these are to one another. Alternatively, if the taxa to which OTUs belong are already known the data can be used to help allocate new OTUs of uncertain affinity to existing classes. The latter techniques form the basis of several matching procedures employed by automated identification systems.

4.3 Hierarchic clustering procedures

The output of hierarchical clustering methods are typically tree diagrams (dendrograms) or their equivalent nested sets with taxa and groups of taxa being connected at different levels according to their overall similarity. It is now widely accepted that their output usually will not reflect evolutionary relationships accurately, such diagrams are best referred to as **phenograms** and they should not be confused with **cladograms** which represent hypotheses about phylogeny. In general, the process of tree construction is sequential but it can be either **divisive** or **agglommerative**. In divisive clustering, splits are sought in the group of taxa that will in some way maximize the collective phenotypic disparity between the two groups formed by the split. In contrast, agglomerative methods start with individual taxa and seek to connect them into pairs etc. in such a way that the similarities between pair or group members are maximized at each level.

The most widely employed of the phenetic clustering methods in taxonomy are **nearest neighbour**, **farthest neighbour**, **UPGMA**, **WPGMA**, and **centroid clustering**. These are discussed in various amounts of detail by Sneath and Sokal (1973) and will only be described briefly here. Perhaps because phenetic techniques have been largely overtaken by cladistic methods in any attempt to relate data to evolutionary history, these techniques are now rarely employed in taxonomy. It should also be noted that the various techniques will quite often lead to substantially different clusters when applied to real data, and from a theoretical point of view phenetic clustering such as the widely used UPGMA model (section 4.3.3) is expected to yield poorer and poorer estimates of phylogeny as the rates of evolution along different evolutionary branches depart more and more from equality. A recent comparison of several of these techniques was provided by Sundberg (1985) who not surprisingly found the resulting clusters to be generally poor at predicting the character states of group members in his sample of nemertean worms.

4.3.1   Nearest neighbour clustering

One of the most widely used procedures is the nearest neighbour clustering method, more correctly known as **single linkage clustering**. The basic data used are normalized intertaxon similarities or distances. Construction of a phenogram using this procedure is illustrated in Figure 4.2. Decisions about whether an OTU will join an existing cluster are based on its maximum similarity to (or minimum distance from) any member of that cluster. The process is initiated by joining the two (or more) most similar OTUs.

4.3.2   Furthest neighbour (complete linkage)

The process of constructing furthest neighbour phenograms is essentially similar to that for nearest neighbour ones except that instead of joining OTUs and groups on the basis of their most similar members, i.e. the shortest distance, they are connected at a level corresponding to the greatest difference between their component members. As with nearest neighbour clustering, the process of joining taxa is initiated by first linking all pairs (or more) that are most similar. This method tends to emphasize differences between clusters.

4.3.3   Unweighted pair-group method using arithmetic averages (UPGMA)

This and the next method are based on joining an OTU to existing clusters on the basis of their average (mean) distance to the members of that cluster. Calculation of the mean distance between two clusters of OTUs is carried out by adding together the distances between all possible pairwise combinations of OTUs in the two clusters and dividing by the number of combinations (Figure 4.3). Apart from this difference in the distance criterion used in deciding which groups will

Similarity matrix

	F	E	D	C	B	A
A	.10	.15	.15	.40	.63	1
B	.08	.19	.15	.53	1	
C	.38	.41	.46	1		
D	.51	.49	1			
E	.60	1				
F	1					

Step	Action	Result	Action	Tree
1	find greatest similarity less than unity	0.63 (A-B)	join A to B at level 0.63	 A B
2	find next highest similarity	0.60 (E-F)	join E to F at level 0.60	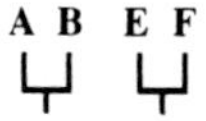 A B E F
3	find next highest similarity	0.53 (B-C)	join C to A+B cluster at level 0.53	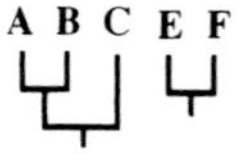 A B C E F
4	find next highest similarity	0.51 (D-F)	join D to E+F cluster at level 0.51	 A B C D E F
5	find next highest similarity	0.49 (D-E)	D already joined to cluster containing E therefore ignore	
6	find next highest similarity	0.46 (C-D)	join A+B+C cluster to D+E+F cluster at level 0.46	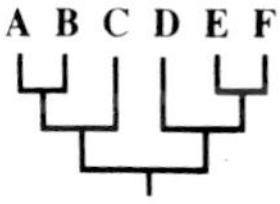 A B C D E F

All taxa linked, tree complete

Figure 4.2 Demonstration of sequence of procedures involved in phenetic clustering of six taxa using the single linkage (nearest neighbour) technique.

join to which others, the tree construction procedure is essentially the same as that employed for nearest neighbour clustering. Because UPGMA uses mean distances it, and WPGMA (see section 4.3.4), is not so subject to the effects of aberrant OTUs and for this reason UPGMA and WPGMA have frequently been preferred over nearest neighbour methods.

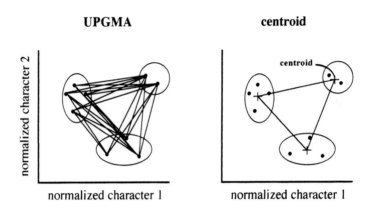

Figure 4.3 Illustration of the difference between the UPGMA and the centroid methods of clustering. In the UPGMA the intercluster distances employed in deciding which clusters will join next is based on the average distance between all possible combinations of pairs of members in different clusters. In centroid clustering, the distance measure is that between the centroids of each cluster.

4.3.4   Weighted pair-group method using arithmetic averages (WPGMA)

This is a set of methods rather than a single one. They are essentially similar to UPGMA except that, as their name implies, they weight distances such that there is a tendency to emphasize separations between larger clusters and they are therefore less subject to the effects of single or small groups of outliers. In practice therefore, the distances are usually weighted by some function of the number of OTUs in each cluster.

4.3.5   Centroid clustering

The basic aim of centroid clustering is similar to UPGMA and WPGMA in that it is designed to reduce the effects of aberrant OTUs but unlike these methods it requires the actual taxa versus character raw data, or more precisely normalized data, rather than intertaxon distances. This procedure then makes use of the **Euclidian distances** (see Figure 4.1) within the character hyperspace between the **centroids** of clusters of OTUs. The process has the tendency to mask some smaller groups because their centroids can be close to the centroid of another cluster with a rather different distribution within the character hyperspace. Sneath and Sokal (1973) briefly describe weighted and unweighted variants.

4.4 Ordination methods

Nowadays, the most useful of the numerical, non-cladistic methods in taxonomy are not so much concerned with arranging taxa into hierarchies but in either finding natural splits between sets of taxa or in finding the best ways of distinguishing similar taxa from one another. The difference between these two aims is one of whether or not it is already known to which groups each taxon belongs. In the former case, a researcher may have at hand a number of specimens but is not sure how many taxa are represented in the sample. In this case, what is wanted is a technique that will show whether the specimens represent an homogeneous sample from a single population or whether they represent two or more distinctly different groups. In the second situation, a worker may know (or suspect) how many distinct classes there are in the sample but, because character states for individual characters intergrade, need a way of reliably distinguishing between taxa on the basis of variation in the whole suite of characters.

Ordination analysis is an extremely large area of mathematical endeavour and numerous books describing methods in detail can be found in most libraries. In biology, ordination has played a large role in ecology and particularly in plant community analysis. Only a brief summary of some of the methods most commonly applied in taxonomy is therefore given here. An easy-to-read but slightly more detailed summary is provided by Alderson (1985) and many methods are also dealt with by Sokal and Sneath (1973).

4.4.1 *Principal components analysis*

Principal components analysis or **PCA** for short, is a widely employed multivariate analysis technique that can reveal hidden groupings or splits in sets of taxa based on two or more sets of measurements. The general principles are best illustrated for just two characters, but in practice there is no limit to the number of characters that can be included in PCA. The basic concept is that the variation in any one original variable (raw data) may not reveal the true diversity of the sample population. Thus in Figure 4.4a while both the x and y variables show some indication of heterogeneity, neither show the sample as comprising two discrete populations. However, brief inspection of the graph of x against y (the original variables) clearly shows that two populations are present. PCA is a method that identifies different axes (in this case other than x and y) that maximize the variance in the data as a whole. Thus in the figured example, two new axes, the first and second principal components, can be found and plotted such that plotting the data along the first principal component shows two clearly separated groups of data points (Figure 4.4b). Like the original variables x and y, the new variable axes are at right angles and are therefore independent of one another. While in this example the original data set comprises only two variables and simple visual inspection can easily reveal the inherent heterogeneity in the data, many situations are more complex, potentially involving many variables, and so plotting any two or three variables against one another may be inadequate to reveal underlying structure in the data. In this case PCA can be extremely helpful.

The actual PCA algorithms, which involve fairly complex matrix algebra and the identification of eigenvectors are too complex to describe here but the principles are relatively easy to comprehend. The aim of PCA is to discover a series of new vectors that are all at right angles to one another (i.e. they are orthogonal) such that the normalized data have maximum variance along the first principal component axis. The second principal component is then selected such that it is at right angles to the first and explains the maximum amount of the remaining variance in the data, i.e. that which is not explained by the first principal component. The third principal component is likewise at right angles to both the first and second and is selected so as to explain the maximum amount of variance in the data not yet explained by the first two principal components, and so on. In practice, the first two or three principal components usually explain most of the useful variation in the data and it is not common for workers to consider the fourth and higher components, although the total number of components that can be identified for a given data set is equal to the original number of characters. It should also be noted that the first principal component is often particularly strongly influenced by OTU

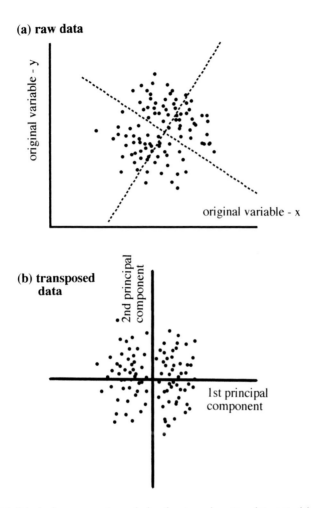

Figure 4.4 Principal components analysis of a two character data set. (a) Data and principal components plotted against original variables; (b) same data replotted with respect to the two principal components. Note that the distribution of data points along the first principal is distinctly bimodal whereas no bimodality is apparent in the distributions with respect to either of the original variables.

size (even after normalization) because many characters display some degree of allometric growth. Therefore, if size is likely to be problematic, the second and third factors may be more informative from a taxonomic point of view.

4.4.2   *Principal coordinate analysis*

This technique is analogous to principal components analysis except that instead of using the raw character state data it employs inter-OTU distances. Thus its output will depend on the nature and behaviour of the particular distance measures employed. It also means, however, that it can be applied to types of data that only come as distances, for example, most immunological techniques (chapter 9) or DNA–DNA hybridization data (chapter 11), whereas principal components analysis cannot.

4.4.3   *Canonical variate analysis*

This technique, which is sometimes also referred to as multiple discrimination analysis, unlike the above two techniques is a discriminant technique which requires the taxa to which OTUs are assigned to be specified beforehand, except of course for any unknown ones that the analysis is designed to help sort out. The analysis makes use of a taxon to taxon distance measure called the Mahalanobis D^2 statistic (Sneath and Sokal, 1973). In essence the procedure involved is similar to PCA in that a set of orthogonal (perpendicular, independent) vectors are sought. The criterion used is that the first vector maximizes the distance between the means of the distances between taxa and the subsequent ones progressively maximize the same value within the confine that they have to be perpendicular to the preceding vectors. This is probably one of the most powerful phenetic techniques available for identification of an unknown specimen. However, as pointed out by Wiley (1981) the fact that a discriminant analysis may be able to distinguish between two sets of OTUs is not a sufficient reason to assume that the two groups are in fact distinct.

Discriminant techniques differ from the principal components and principal coordinates analyses in that the total number of discriminant axes they can find is one fewer than the number of groups of OTUs.

4.4.4   *Non-metric multidimensional scaling*

This last technique is rather different from the preceding three in that it is based on the notion of how well a taxon known from n characters can be represented in dimensions less than n. Its raw data is an ordered list of distances (or similarities) and it can even utilize a simple rank order of distances. From this the user decides on a new number of dimensions, *fewer* than the original set of characters, and as this has to be observed it is usual to choose either two or three dimensions. Whether the new ordination is a fair reflection of the original higher dimensional data set is determined by a measure called the stress which compares the original rank order of OTU–OTU distances with those in the transformed data. The lower the stress the better the correspondence. One of this technique's big advantages is that it readily allows for missing data.

CHAPTER FIVE

KEYS AND IDENTIFICATION

> All biologists are involved in identification . . . but the taxonomist is especially concerned, since he produces or alters the classifications of organisms as well as providing the tools by which his fellow-scientists can identify their specimens.
>
> Walters (1975)

5.1 Introduction

An often underestimated role of taxonomy is that it not only provides names and classifications for organisms, it also provides the means for identifying them. In fact, for many users of taxonomy this may be its most important function.

As the number of known taxa has continued to increase at pace ever since the days of Linnaeus, so too has the need for better methods to enable accurate, quick and easy identification of organisms. In principle it ought to be possible to compare any specimen with each relevant original description and to accept as an identification the one which gives the best match, but alas the original descriptions of the members of any taxonomic group are usually widely distributed across many frequently obscure literature sources, written in inconsistent style and language over a period of time when concepts may well have changed considerably, and they frequently include a level of redundancy due to the same taxon having been described as new on more than one occasion. Thus, to go about identification in this way would be time-consuming and difficult, not to mention costly and unreliable. One of the most important roles of taxonomists is therefore to carry out revisionary studies of groups, usually just called **revisions**, in which **synonyms** are identified, uniform sets of characters found that can be used to recognize the taxa involved and, of course, to provide more practical means of identification.

For most groups of organisms the most popular and successful means of identification has been through the use of **keys** and, in particular, the **dichotomous key**. However, this is by no means the only identification system and for some taxa, such as many microorganisms, identification may be achieved better or at least more practically using alternative methods especially matching procedures. Much of this chapter will be devoted to keys, their properties and some general considerations on how best to construct them. Matching procedures will be described briefly later in the chapter and some of the coefficients employed are described in chapter 4.

As its name implies, the dichotomous key consists of a series of divisions or dichotomies each of which presents two alternative and exclusive sets of characters that distinguish one group of taxa from another. Working through the key progressively narrows the number of possible identifications until at the end only a single taxon is left. The alternatives which make up each dichotomy in the key are called **couplets** and within a couplet each of the two alternative sets of characters is often referred to as a **lead** (or **leg**).

5.1.1 *Purpose of keys*

The purpose of a key is to enable identification, and it should not be a vehicle for expressing systematic opinions. This is especially so for two reasons. First, good systematic characters are very often poor or even unusable key characters, and second, classifications are all too frequently subject to modification. Thus, although it may sometimes be possible to construct a good key that reflects a perceived natural classification, this is seldom the case and certainly should not be a priority objective. Unfortunately, classifications are sometimes presented in the form of a **synoptic key** but in these cases they should never be relied on for identification purposes.

5.1.2 *Good practice in writing keys*

Keys come intended for a wide range of potential users with concomitant differences in experience and requirements. Some are aimed at giving only a crude level of identification, others take the user to species or subspecies level. Further, keys may be constructed for generally well understood groups, such as the butterflies of Great Britain, while others will be the result of original revisionary studies of poorly known taxa. Despite these immense differences in aims and usage, several factors can be identified as being important for the key to be as effective as possible. Adherence to 'good practice' is paramount as there is little point in producing a key that people cannot use just as there is no point in producing a key that will not work. Many authors have published on good practice in key construction and writing, for the most part with a fairly good consensus, and only a few general points will be made here. Many important points are discussed at greater length by Tilling (1984, 1987).

Ideally new keys to a group, especially little-known ones, should be produced by an expert who has studied all available authenticated specimens and as much additional material as is realistic. This also ought to mean undertaking proper revision of the group with consultation of all original descriptions and all available type specimens.

Views differ considerably between taxonomists as to how many characters should be included in a couplet. Too many may well cause confusion and the user might then lose confidence in the key or, worse still, make a wrong decision. However, **monothetic keys**, that is ones that employ only one character per lead, can be problematical too. A single character will often be too few because the user can easily be prevented from progressing further through the key should the specimen being keyed have the relevant structure obscured or missing. This point is especially relevant to characters that are particularly vulnerable to damage such as an insect's antenna, and it also follows that monothetic couplets should be particularly avoided near the beginning of a key if there is a real possibility of the structure concerned being broken or obscured. The best compromise is probably to use two or three characters in any given lead, assuming that is that more than one reliable character for a given split exists.

Given that keys should enable rapid and accurate identification, it would be misleading not to point out that many users experience difficulties particularly when using any given key for the first few times. This results from a number of causes. One common problem is ambiguous wording which leads to uncertainty about the meaning of a couplet. Another common problem is the difficulty that many users (apart from the author of the key of course) may have in appreciating what are often subtle character differences. Indeed it should always be realized that for most characters either experience may be needed in recognizing the alternatives or confirmation is needed that the user has indeed made a correct identification (or otherwise).

Good, clear and copious illustrations, especially when provided with arrows highlighting the key points, are an enormous help and should be regarded as an essential rather than an optional extra; sadly figures are absent or rare in many older works. Most importantly, key characters should be truly contrasting with no possibility of overlap. Hall (1970) recommends that to be usable, the differences between leads involving continuous characters should be greater than 0.1 of the total range of the character. Finally, it should be noted that difficulties in using a key not infrequently result from an initial misidentification of the group to which the organism belongs and thence to the use of an inappropriate key.

The usability of a key also depends on clear writing, and can be greatly enhanced by following some simple stylistic guidelines. Style should be consistently telegraphic, that is short without unnecessary linking words and articles such as 'a' and 'the'. However, it should not be so telegraphic as to obscure the meaning. Within a couplet, leads should start with the same word and leads of adjacent couplets should start with different words, this way the user's eye will tend to fall back on the correct couplet after having examined the specimen at hand.

Providing couplets or leads with labels (in parentheses) indicating the previous couplet or lead number facilitates backtracking if it is thought that the specimen is running to the wrong part of the key. This simple addition can now be achieved readily with computerized key construction techniques and is especially useful in the case of **indented keys** (see section 5.2) in which corresponding leads can be widely separated.

Supplementary characters are also often a great help if treated with appropriate caution. These are sometimes included within the key itself, usually in parentheses, but are better separated from the diagnostic characters in some more obvious way such as in an indented subparagraph. Supplementary characters are typically invariant in one alternative but variable for another, or show some degree of overlap in character state ranges.

Percentage indications may sometimes be provided to give the user an idea of the likelihood of encountering a given character state or taxon. However, if some taxa following from a particular couplet do regularly display a character, encountering a statement like 'winged (90%) or wingless (10%)' may be confusing to a novice user and could even lead to a misidentification. Similarly, statements such as 'usually hairy' are of limited value in a key and are best avoided.

Exception clauses, e.g. 'leaves hairy except in *glabrescens*', should be avoided as such devices require the user to be able to distinguish *'glabrescens'*. It is far better to key the exception separately as even if enough supplementary information is given to enable this taxon to be recognized, the amount of extra character checking that might be required could easily lead to confusion. Similarly, long keys especially tend to resort to complex constructions with lots of qualifiers such as 'however' or 'although' and alternatives such as 'either' and 'or', and 'if' and 'then' clauses. Such devices can be totally logical but unless very carefully executed, they all too often cause confusion. For many purposes, use of these structures is counterproductive especially in keys intended for use by non-experts (Tilling, 1987). It should be recognized that use of these awkward grammatical constructions typically results from either trying to make a key do two things, i.e. to identify taxa and simultaneously to preserve some concept of natural or systematic groupings, or from trying to key each taxon out only once when in reality the variation within taxa may make it far more sensible to key a taxon out on multiple occasions. An alternative system can usually be constructed which reads more easily; specifically, keys can contain loops (**reticulate keys**; Osborne, 1963). Although reticulate keys may not always improve overall accuracy for reasons discussed below, they can greatly improve the confidence a user has in using it.

It is generally held that keys should be strictly dichotomous (i.e. have only two options available at any one decision point/couplet) as the greater choice of a polychotomy will make decision making harder. This is generally true though there are instances where strict adherence to dichotomy would be superfluous, for example, if a multistate character is simple and unambiguous such as the number of eyes possessed by springtails.

Finally, wherever possible, keys should be tested by others before they are committed to publication. Such testing may not be so practicable for revisionary works but without doubt it is one of the easiest ways of discerning poor wording, inconsistencies and errors (Tilling, 1987).

5.2 Types of keys

Keys can be broadly divided into two groups, single entry dichotomous keys and multiple entry keys or **polyclades**. They differ in that in the former the order of questions that need answering to identify a given taxon are prescribed from the outset whereas in multiple entry keys the user at each stage has the choice of selecting a particular character for consideration and therefore the sequence of character analysis and the route through the key is largely dependent on the user.

5.2.1 *Dichotomous keys*

Over the years several numbering systems and other stylistic conventions have been employed to direct the user from one question (lead) to another. In practice only two basic arrangements are in common use, the bracketed (or parallel) key and the so-called indented (or yoked) key. Some computer packages also output tabular keys of various forms. In the one produced by Dallwitz's KEY program (Dallwitz, 1974; see example below) the identification structure and the text are presented separately; the ability of these to represent the decision-making process may eventually lead to these gaining in popularity. The following example keys are arranged from Brown's work on British grasshoppers and allies (Brown, 1983) and will serve to illustrate the differences.

Bracketed key

1(0) Fore legs not modified for digging, similar to mid legs 2
       Fore legs highly modified for digging .... Gryllotalpidae (mole crickets)
2(1) Antennae shorter than body, rather stout 3
       Antennae longer than body, slender 4
3(2) Pronotum saddle-shaped, not extending over abdomen .... Acrididae
                                                  (grasshoppers)
       Pronotum posteriorly produced, extending far back over abdomen ..... Tetrigidae
                                               (groundhoppers)
4(2) Tarsi with 4 segments .... Tettigoniidae (bush crickets)
       Tarsi with 3 segments .... Gryllidae (crickets)

Indented key

1(2) Fore legs highly modified for digging .... Gryllotalpidae (mole crickets)
2(1) Fore legs not modified for digging, similar to mid legs
3(6) Antennae shorter than body, rather stout
4(5) Pronotum saddle-shaped, not extending over abdomen .... Acrididae
                                                  (grasshoppers)
5(4) Pronotum posteriorly produced, extending far back over abdomen ..... Tetrigidae
                                               (groundhoppers)
6(3) Antennae longer than body, slender
7(8) Tarsi with 4 segments .... Tettigoniidae (bush crickets)
8(7) Tarsi with 3 segments .... Gryllidae (crickets)

Tabular key

ACRIDIDAE	3A	1A	2A
TETRIGIDAE	3A	1A	2B
TETTIGONIIDAE	3A	1B	4A
GRYLLIDAE	3A	1B	4B
GRYLLOTALPIDAE	3B		

Characters

1A   Antennae shorter than body, rather stout
1B   Antennae longer than body, slender

2A   Pronotum saddle-shaped, not extending over abdomen
2B   Pronotum posteriorly produced, extending far back over abdomen

3A   Fore legs not modified for digging, similar to mid legs
3B   Fore legs highly modified for digging

4A   Tarsi with 4 segments
4B   Tarsi with 3 segments

Another form of 'tabular key' was described by Newell (1970) and involves the simultaneous appraisal of a small set of characters at each level. Although not strictly a dichotomous key and not in wide usage, this system often allows the identification of incomplete specimens as it provides information on all characters for all taxa at a given separation level.

5.2.2   *Multiple-entry keys*

Multiple-entry keys (now often referred to as polyclades) enable the user to select for examination those characters which seem most obvious or, at least, are present in the specimen at hand. In this respect they are more powerful than standard dichotomous keys in which the sequence of character examination is prescribed from the outset. They can be particularly useful when a specimen to be identified is incomplete or, for example with botanical material which according to time of year might be purely vegetative, in bud, in flower or in seed.

Multiple-entry keys have until quite recently been most commonly presented in the form of punched cards. Indeed, these may still have a role for some identifications in field situations and their use has been particularly popular with botanists. Two variants of these card systems are found, so-called peek-a-boo (or centre-punched card) sets and edge-punched cards.

In the peek-a-boo system, each card represents a character and set locations on each card represent taxa. Taxa displaying a character have a hole punched in the card at a constant position from card to card. Cards are selected if the specimen displays the character they represent and cards thus selected are stacked together and selection continues until only one unobscured hole remains in the aligned pack, this hole corresponding to the identified taxon. In the edge-punched system each card represents a taxon and the holes at the edges the characters. If a taxon displays a character the hole is left intact otherwise it is clipped open. Thus by inserting a long needle along a hole through the pack of cards, those cards with intact holes (representing taxa with a character) can be separated from the remainder. Again, repeating this process with user-selected characters will ultimately leave one card and hence an identification is made.

Although these systems give the user flexibility, they are not intelligent systems in that the user may have to examine many more characters than is strictly necessary. This feature can be readily overcome by computerization which can allow the user to select which character to examine next while the program offers advice about which ones would be most likely to lead to a rapid identification.

5.3 Efficiency

The efficiency of a key is determined by a number of factors ranging from the ease with which characters can be examined and the reliability of their interpretation, to the number of questions that need to be answered to identify an average taxon (Hall, 1970).

5.3.1 *Length of dichotomous key*

The length (L) of a key can be defined as the average number of couplets that have to be used in order to identify a taxon (Osborne, 1963; Dallwitz, 1974). Length is determined by the topological arrangement of the key (Figure 5.1). With T terminal branches (taxa if each taxon is keyed only once), the comb-like arrangement of Figure

5.1 (*left*), representing the longest possible key, has a length of $(T+1)/2 - 1/T$, whereas the fan arrangement (Figure 5.1 (*right*)) representing the shortest possible tree, has $L \sim \log_2(T)$. As one of these is approximately a linear function of T and the other a logarithmic one, the relative difference between them increases with T. For example, with $T = 8$, $L_{shortest} = 3$ and $L_{longest} = 4.375$, whereas with $T = 128$, $L_{shortest} = 7$ but $L_{longest} = 64.508$. In practice, the advantages obtained through maximizing key symmetry (evenness of splits) only become important when more than 10 taxa are involved (Figure 5.2; Hall, 1970).

5.3.2   Reliability

Another important implication of a key's architecture is its reliability. In the real world a key is unlikely to be 100% accurate, each question having a finite probability of being wrongly answered due to some combination of constructor error, user error or specimen ambiguity. Osborne (1963) investigated key reliability (R), defined as the average probability of making a correct identification, and showed a strong dependence on average key length such that shorter keys should be preferred, and this becomes especially important as the characters employed become less reliable. Interestingly, however, Osborne also showed that there are exceptions in that for some topologies a longer key can be more reliable than a shorter one. The difference in the probability of making a wrong identification, W, between the shortest key, W_{short}, and the longest, W_{long}, is dependent on the number of taxa, and is virtually independent of average character reliability, being approximated by

$$W_{short}/W_{long} \sim (2/T)\log_2 T$$

Thus with only eight taxa the shortest key is only about 75% as likely to lead to a wrong identification as the longest one, whereas with 32 taxa the long form of the key is approximately three times as likely to give a wrong identification.

5.3.3   Choice of characters

Two factors have a strong influence on the choice of characters to be included in a key: their observability and their reliability. Unfortunately, the two properties are often poorly correlated. In general, it

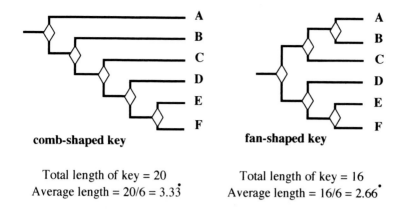

comb-shaped key   **fan-shaped key**

Total length of key = 20   Total length of key = 16
Average length = 20/6 = 3.33̇   Average length = 16/6 = 2.66̇

Figure 5.1 Topologies of two extreme forms of dichotomous keys for six taxa (A to F). So-called comb-shaped or pectinate key (*left*); so-called fan-shaped or symmetric key (*right*). The total length of the key is the number of couplets that have to be consulted if each taxon is to be identified once. The average length is the total length divided by the number of taxa.

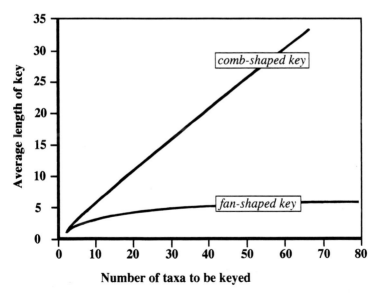

Figure 5.2 Relationship between average length of a dichotomous key and the number of taxa included for the two extreme key topologies (each taxon included in key once only) (see Figure 5.1).

is good practice to arrange keys such that couplets leading to splits between major groups of taxa and those near the beginning of the key make use of the best characters in terms of observability and reliability. This way any mistakes or impasses are likely to be less misleading and less troublesome.

5.3.4 *Likelihood of encountering taxon*

Users unfamiliar with a group often become worried when the specimens they are trying to identify seem always to run almost to the end of a lengthy key, fearful that they have made a mistake somewhere before. A likely cause of this is that the taxa which key out at the end of keys are difficult to characterize and so the writer has repeatedly removed more easily recognized ones earlier, thus leaving the 'dregs' to the end. Unfortunately, all too often the dregs include large, important and frequently collected taxa.

5.4 Computerized key construction

Constructing a good key is no easy task even without considering such factors as length and reliability. Modern computer packages, however, offer the writer the opportunity of trying out many different topologies from which an appropriate selection can be made, whereas to do so by hand could be immensely time-consuming. One of the strongest assets of an automated process is the ease with which either new taxa can be added to a key (hand renumbering is a most tedious and potentially error-prone task) or specialist keys (e.g. to species in a particular region or parasites of a particular host) can be constructed.

Computerization of key construction requires at its simplest, a taxon × character data matrix and in early applications this was usually the input format (e.g. KEYGEN by Pankhurst, 1971). However, such formats are not readily amenable to making changes and as an initial solution to this problem Watson and Milne (1972) wrote a front program, EDITOR, to overcome these difficulties. More recently, Dallwitz (1974) has introduced a free format data entry system to go with his key construction program, KEY. Dallwitz's system, DELTA (DEscription Language for TAxonomy) is now widely used and has been adopted by Pankhurst in more recent packages.

All key construction programs work by scanning all possible characters at each decision level and determining which one best fits the programs' prescribed selection algorithm, which also usually incorporates user-defined parameters concerning length, character weighting and taxon abundance. One of the simplest was used by Rypka and Babb (1970) in which the selection parameter, S, was determined solely by the equality of the subgroups that would be formed if that character was used, thus minimizing key length. In that case $S = n_1 n_2$ where n_1 and n_2 represent the numbers of taxa that would comprise the two subgroups formed by the split. Thus Rypka and Babb's S could range from 1 to $T^2/4$ where T is the total number of taxa. Since then various other authors have advocated many modifications to this system aimed at changing the precise dynamics of this selection parameter. Pankhurst (1970), for example, advocated as an evenness-of-split function to be minimized,

$$|n_1 - T_r/2| + |n_2 - T_r/2|$$

where T_r is the number of taxa remaining to be keyed-out, i.e. $n_1 + n_2$. While other functions weighting even more heavily in favour of evenness-of-split have been suggested, Wilson and Partridge (1986) suggest that to give too much weight to this criterion is probably unwise and they conclude that further subtle variations on the weighting function are rather pointless.

Computerized construction packages such as KEY (Dallwitz and Paine, 1986; Partridge *et al.*, 1988) usually also incorporate weighting functions that allow characters to be weighted to allow for ease of appreciation of their reliability, and taxa to be weighted in accordance with their relative abundances so that commonly encountered taxa will be keyed-out earlier. This flexibility permits quick production of a number of alternative keys from the same data set. From these the one that most fits the bill in terms of ease of use, reliability, and other criteria can then be carefully selected. An alternative offered by a program described by Pankhurst (1988a) is interactive computerized key construction (not to be confused with interactive identification) in which the key constructor is offered the computer's choice of best character at each stage but this choice can be overridden by the constructor if necessary on the grounds of experience.

Numerous computer-based key construction packages have been developed over the years, both for dichotomous key construction (Hall, 1970 – BOLAID; Pankhurst, 1970, 1971 – KEYGEN; Dallwitz,

1974 – KEY; Ross – CLASP; Payne, 1975 – GENKEY) and polyclade (multiple-entry key) construction (Morse – CARDKEY; Pankhurst, Schulz – CDKEY; Pankhurst and Aitchison, 1975 – POLPUN; White, in prep. – CABIKEY).

5.4.1 *Interactive identification*

Computerization is now revolutionizing the use of multiple-entry keys (polyclades) as the ability of computers to search rapidly through data sets makes them ideally suited to this mode of identification. Several good programs are now available, some of which take user-defined data sets while others are circulated with predefined data for identifying particular groups of taxa. Computerized systems are particularly useful for identifying large and complicated taxonomic groups and they may become widely used for many specialized applications such as crop pests or diseases. Recent programs generally have a number of user-friendly requirements such as providing lists of remaining taxa, lists of best characters, confirmatory characters, minimum sets of diagnostic characters, taxon descriptions, lists of most easily confused taxa and backtracking facilities, though not every package offers all of the above.

For maximum reliability from a computerized polyclade, the user should only answer questions that they know the answer to with certainty, for it is the flexibility of search path that makes these systems so much better than a predefined printed key. Some programs keep a tally of how often answers have been given to a particular question and may progressively weight against the use of questions that are unpopular (Wilson and Partridge, 1986).

Computerized polyclades can relatively easily make allowance for taxa that are incompletely known. Efficiency of the search path can also be improved by the use of associated characters such that if the specimen being identified lacks some structure (e.g. flowers) then the program will restrict questioning to non-floral characters. Further, if a question about one structure has been answered successfully then the subsequent question may also refer to the same region of the specimen because the user is already familiar with it. Another potential improvement over written keys, though perhaps a more contentious one, is the ability to bias identifications towards common taxa if ambiguity occurs. Whether this is desirable or not is probably more a function of the user rather than the taxa being keyed.

Particularly for some large computerized keys, it is also desirable to allow for some degree of error in the answers provided. This enters the realm of the expert system, an approach to taxonomic identification that has been embraced by a number of workers (Pankhurst, 1970; Atkinson and Gammerman, 1987). With expert systems a particular answer does not necessarily exclude any taxon from further consideration because it is always assumed that questions have less than a 100% chance of being answered correctly. The role of the expert system is particularly relevant to identifications by non-experts whose decisions may be more or less error prone (that is not to say that human experts do not themselves make mistakes from time to time). Thus for an analogy, a computer running an expert system program is acting rather like a human expert trying to carry out an identification over the telephone. The expert asks questions but treats each individual answer with some skepticism and on the basis of years of experience modifies subsequent questions if the answers being received seem incongruent with expectations.

Expert systems such as SYSEX by Gammerman *et al.* and EXPERT KEY by Atkinson and Gammerman (1987) employ probabilistic methods explicitly. Some other expert systems, e.g. ONLINEKEY of Wilson and Partridge (1986) also include a **heuristic** (self-training) system that allows for modifications of the initial writer-defined probabilities based on results obtained during the program's subsequent usage. A recent development that is proving of growing importance is the use of neural networks as a self-training identification system.

Great improvements in computerized identification systems have taken place and are still underway, but it would be wrong to give the impression that they are sweeping away old-fashioned printed works. For many applications it may still be simpler to use a printed key rather than to enter data into a terminal, although there are some areas in which exceptions are more common including some essentially laboratory-based situations, such as the identification of pathogenic microorganisms. Further, whilst it is now commonplace in the developed world to have computers in every laboratory, this is not yet true for much of the developing world where the spread of computers has been rather slower. However, in the developing world, computers may in fact be more readily accessible than good library facilities, making interactive identification potentially far more important there than in the developed world. At present the vast majority of keys are still only available in printed form and until a far larger number are available for computers, their usage will necessarily be supplementary for most users. Perhaps the needs of the developing world for accurate up-to-date means of identification will provide some stimulus for an increase in production of computerized keys and expert systems, particularly in the field of pests and diseases.

Probably the major disadvantage with computerized identification until recently has been the difficulty of storing high quality images so essential for good keys. However, this restriction is rapidly vanishing both as new high density storage media such as compact discs are developed and as the memories of personal computers increase in leaps and bounds. Likewise, increased computer power makes possible far more user-friendly applications with pull-down menus, etc. which will obviously help to attract potential users.

5.4.2 *Matching*

Not all groups of organisms or circumstances lend themselves to identification using standard format keys. It may be difficult, for example, to examine all key characters required, or there may be considerable intrinsic variation between individuals. In such cases, the use of a diagnostic key might at best be difficult or at worst might be impossible. Further problems can arise, for example, in the identification of bacteria and Protozoa, which often rely on a series of biochemical tests some of which might give ambiguous results. Thus, in the field of pathology, identification has tended to use matching algorithms to find the best fit between the available information on a pathogen and information on strains and taxa stored in a reference database.

5.4.3 *Automated taxon descriptions*

Having encoded characters for each taxon, for example in the DELTA format, a computer package such as Dallwitz and Paine's (1986) CONFOR can create natural language descriptions of all taxa from the DELTA input data. However, the usefulness of these as they stand is questionable as they are just a longhand version of the condensed input. Further, there may be subtle differences between taxa that should be included in a description but that would not make good key characters and may therefore not have been included originally. The descriptions that are output should therefore be considered as a framework rather than a completed product.

It can be argued that descriptions in the traditional sense have little real value and that only a **diagnosis** listing the critical points by which a new taxon can be distinguished from others is needed. At first this may seem a strange point of view, but in practice taxonomists are virtually never able to rely on original descriptions in order to make a certain identification of a specimen. After all, if they could, then they would virtually never need to examine type specimens. The truth is that taxonomists cannot predict what features will be necessary to separate their taxa from those yet to be discovered. The contrary view espoused, for example by Hawksworth, is that descriptions ought to be as complete as possible (e.g. Sigler and Hawksworth, 1987). Thus the best policy is probably to restrict descriptions to those characters that differ within the group, family, genus or whatever, under study.

5.4.4 *Databases*

With current advances in computerization, worldwide access to vast amounts of data is now a reality and one that taxonomists have not been slow to realize. A considerable while ago, workers like Pankhurst, and Dallwitz and Paine were developing standardized taxonomic data structures, such as the DELTA format, with provisions for incorporating additional information. The general desirable features of a taxonomic database have been summarized by Pankhurst (1988b). In addition to their physical characters, these include for each taxon bibliographic references, distributions, ecological data, economic or medical importance and nomenclatural history so that references to synonyms can be identified. The utility of such databases derives from the ability to extract all relevant information on a number of separate fields. For example, a floristic database could be used to construct keys to the floras of particular regions such as national parks, or perhaps a list of poisonous plants occurring in a particular habitat.

It should be remembered that memory considerations no longer pose a real limit on the amount of data that can be stored and quickly accessed and therefore databases can incorporate not just citations but the complete references for a given taxon, including illustrations. Even more can be achieved with computer networks. Perhaps one of the most significant of recent developments is the idea that it is now possible to computerize all specimens in a museum (even all museums). To this end a number of institutions are barcoding their new specimens so that complete data on them can be stored electronically. This even extends as far as using satellite positioning equipment precisely to locate the origins of all material. This wealth of information will eventually constitute a huge data set, and it will then be possible to interrogate it to find the answers to many questions that it would be impractical to attempt to answer at present. For example, a conservation agency would find it very difficult today to get access to all relevant information on the insect fauna of a site before and after a pollution incident even though this information might theoretically exist in the form of collected specimens distributed amongst potentially many insect collections. To sort through specimens by hand for label data would be a hopeless or, at least, enormously costly enterprise, with no guaranteed benefit. However, if such individual specimen data were all available on a computer network, it would be a relatively simple task to assemble the necessary species lists. Thus there can be little doubt that for at least some groups and some tasks, databases will become increasingly valuable tools in both taxonomic research and conservation.

CHAPTER SIX
NOMENCLATURE AND CLASSIFICATION

In any taxonomic work it is the sole responsibility of the author(s) to judge from the evidence obtained whether taxa are to be (a) described as new species, (b) considered as valid species, or (c) put in synonymy with recognised species. Later studies might then reveal the correctness of the systematics of the group studied.

Jelnes (1986)

6.1 Introduction

Nomenclature, that aspect of taxonomy that deals specifically with the naming of organisms, is perhaps its least-loved aspect, although without it biological science would certainly be in a far poorer state. While nomenclature is littered with pedantry and what often seems to be a wealth of trivia, it forms the basis by which scientists can name and cross-refer to organisms. Without names, after all, where would biology be. In general the **Codes** that govern in a gentlemanly way the actions of taxonomists as they affect names, try to be logical in rather the same way as the rules of cricket are logical. However, logic does not always coincide with simplicity. Added to this is another source of contention, namely that the rules governing nomenclature owe a lot to historical accident. Few would doubt that if we could wipe the slate clean, a far more coherent set of rules could be obtained. Unfortunately, for most groups far too much has been published to effect a change now and so taxonomists and the users of taxonomy have to learn the relevant rules and to live with them. Probably the one major annoyance for any general student of taxonomy is that for historical reasons, the nomenclatures of different major groups of organisms have come to be subject to different sets of rules and while it would be nice to have to learn just one set, this is a long way off yet (Ride, 1988).

What follows cannot hope to cover more than a small part of the voluminous and intricate rules of nomenclature as they apply to just one kingdom of organisms, let alone all living organisms. Therefore, an outline of the major principles will be presented and a few of the differences between codes that might be particularly relevant to the needs of the students highlighted. A more thorough survey is provided by Jeffrey (1989).

6.2 The binomial system and the hierarchy of taxa

While taxonomy may have begun with Aristotle, modern taxonomy has an accepted beginning in much more recent times. In fact, modern taxonomy can be said to start in 1753 with the publication of the first part of the *Systema* by the Swedish naturalist Linnaeus, though for some groups of organisms, later starting dates are specified by the relevant Codes (see section 6.6 and 6.8). In 1757, Linnaeus was ennobled which resulted in a name change to Karl von Linné by which he is also known. Linnaeus proposed a system of nomenclature in which every species of organism could be assigned a name comprising two parts, a specific name which is typically descriptive of the species, and a generic name which specifies a group to which that species belongs. A species is specified by the combination of both its specific and generic names and thus, because it requires two names, this is referred to as the **binomial** system. Despite the occasional proposals of alternative systems, the Linnaean binomial system has stood the test of time and is so firmly embedded in biology now that it is almost impossible to imagine its overthrow.

6.3 The International Commissions

At various times biologists concerned with the taxonomy of particular major groups have come together at international meetings to discuss the plethora of problems associated with nomenclature. As a result of such meetings, a set of International Commissions or Committees have been created to lay down sets of rules and recommendations covering the application of nomenclatural procedures to their particular groups of organisms. Of long standing are commissions dealing with zoological and botanical nomenclature though the names of the organizations concerned have changed from time to time. In addition there are bodies dealing with the nomenclature of cultivated plants and bacteria, and to these the International Committee on Nomenclature of Viruses founded in 1966 (now however, called the International Committee on Taxonomy of Viruses, **ICTV**) and the International Commission on the Taxonomy of Fungi (**ICTF**) have recently been added (Sigler and Hawksworth, 1986).

6.3.1 *Codes of nomenclature*

For each major group of organisms an international commission is responsible for defining the rules and procedures that must be employed for the application of scientific names in the group as well as various recommendations for good practice. For animals, including the animal-like Protista (Protozoa), the relevant publication is the *International Code of Zoological Nomenclature* (**ICZN**) (International Commission of Zoological Nomenclature, 1985), for algae, plants and fungi (including their fossils but excluding cultivars) the rules are set out in the *International Code of Botanical Nomenclature* (**ICBN**) (International Botanical Congress, 1983), for cultivated plants special provisions are given in the *International Code of Nomenclature for Cultivated Plants — 1980* (**ICNCV**), for bacteria that are governed by the *International Code of Nomenclature of Bacteria* (**ICNB**) (Lapage *et al.*, 1975) and virus classification and nomenclature is covered by the rules of the International Committee on the Taxonomy of Viruses (**ICTV**) (Wildly, 1971; Fenner, 1976; Matthews, 1979). Recently additional and modified recommendations have been proposed for fungi by Sigler and Hawksworth (1986) under the auspices of the **ICTF** although for the time being at least, fungal nomenclature is intertwined with the rest of botanical nomenclature.

These rules are set out in publications, the '**Codes**', which result from the occasional Congresses of the respective bodies. As such the Codes are subject to changes from time to time, and this has been particularly true of the rules for naming viruses which have still to attain a real consensus and wide acceptance. However, that changes are sometimes necessary even for other groups should not come as a surprise. For example, new and unforeseen situations may arise and the needs of taxonomists and users of taxonomy are not necessarily a constant. It would have been virtually impossible for earlier nomenclaturists to have anticipated the shear diversity and complexity of the problems that can arise.

The fact that there are different Codes for different groups has historical origins but there are some general differences between the groups that may predispose their respective taxonomists to have different nomenclatorial needs. For example, hybridization in animals is not particularly common whereas in plants it occurs more frequently, and importantly it quite often produces viable fertile new taxa. Another significant factor is the difference in reproductive mode between the prokaryotes and the eukaryotes which affects, for example, gene flow and hence species definitions.

6.3.2   *Independence of the Codes*

For most groups, the Codes of nomenclature are effectively independent of one another, and therefore names of taxa governed by one code do not generally enter into questions of **homonymy** with identical names proposed for groups governed by other codes. Thus, while no two genera of animals or two genera of plants can have the same valid name, it is permitted, although not recommended, that an animal can have the same generic name as a plant. For example, the genus *Agathis* is a valid name for both an important genus of southern hemisphere trees and for a large genus of parasitic wasps. Similarly, the generic name *Allotropa* applies to another wasp as well as a parasitic plant. However, while it is permitted for genera and higher taxa of animals and plants or fungi to have the same names, it is a recommendation of the ICZN that generic names of animals should not be the same as those used in any other group, and the ICNB forbids a bacterial genus from being the same as the name of a fungus, protozoan or alga. Further details of restrictions are given in Table 6.1.

6.4   Basic principles of nomenclature

The basic principles of each of the Codes are largely the same. Roughly these can be summarized as:

  1. to provide stability in the naming and classification of organisms;

Table 6.1 Extents of coverage and independence of Codes of nomenclature

Group	Relevant Code	Independence in terms of homonymy
Metazoan animals	ICZN	All other groups except Protozoa
Protozoa	ICZN	All other groups except metazoa (ICZN) and bacteria (ICNB)
Higher plants	ICBN	All other groups except fungi, algae and blue-green algae
Fungi	ICBN	All other groups except higher plants, algae, blue-green algae (ICBN and ICNB) and bacteria (ICNB)
Algae	ICBN	All other groups except higher plants, fungi, blue-green algae (ICBN) and bacteria (ICNB)
Blue-green algae (Cyanobacteria)	ICNB and ICBN	All other groups except higher plants, algae, fungi (ICBN) and bacteria (ICNB)
Bacteria	ICNB	All other groups except for algae, fungi and Protozoa.

2. to ensure that any given taxonomic grouping of a given rank can have only one correct name.

6.4.1 *Priority*

With the exception of the ICTV, where two or more names have been applied to the group in the past, the correct name will be the one that was published on the earliest data, providing it is acceptable in terms of the rest of the Code. This system is referred to as the law of **priority**.

Application of the law of priority will ultimately lead to stability in nomenclature, all other things being equal. However, the use of priority has ironically caused considerable discontent among many of the users of nomenclature because quite often the name with priority (i.e. the senior or oldest name) is not the one in common usage, often because it was originally published in some obscure place. When common taxa widely used in other areas of science, industry, agriculture or medicine are involved, strict application of the law of priority can cause considerable confusion and in the past this has on occasion resulted in a species being referred to by both names for a protracted period, hardly a satisfactory arrangement and potentially a quite dangerous one.

Views differ widely about how far the law of priority should be taken, and the various Codes and Commissions have not been consistent in their treatment. However, both botanical and zoological nomenclature now accept the concept of the **conserved name** or **nomen conservandum**, i.e. a name which does not have priority may nevertheless be conserved as the correct name for a taxon by the explicit use of the **plenary powers** of the relevant International Commission if that body decides that to do so would be in the best interest of taxonomy. This situation was dealt with in one fell swoop for bacteria in 1980 by the publication of the *Approved Lists of Bacterial Names* (Skerman *et al.*, 1980).

6.4.2 *Synonymy*

Ideally, every taxon would only ever be given a single scientific name. Unfortunately, it often happens that two or more different taxonomists working on the same group will independently publish original descriptions of a given taxon resulting in two or more competing names for the same entity, and these names are referred to as **synonyms**. Synonymy may arise in several different ways, for example, for many years hydrozoan nomenclature has been confused as a result of taxonomists working independently on the medusa (jellyfish) and polyp stages of the same organisms, both of which received near complete taxonomic treatments resulting in most species having two names. In other cases, two taxonomists may disagree about the number of taxa involved in a particular group with the result that one will subdivide the classification to a greater extent than the other. In this case, one worker will consider some of the other's names as synonyms. These two types of synonymy differ in that in the first case there is no doubt that a particular entity (i.e. a species of hydrozoan) has received two names, while in the second case there is an element of subjectivity about what comprises a single entity. When a synonymy is dependent on the opinion of a taxonomist concerning the identity of a taxon it is referred to as a subjective synonym.

No matter how or for what reason a taxon has received more than one scientific name, it is a fundamental rule of all the Codes of nomenclature that all taxa can have only one **valid name**, that is only one of the names that have been proposed can be regarded as the correct name, and all others are invalid. In general, the first name that was proposed for a taxon, the senior synonym, will be the valid name.

6.4.3 *Homonymy*

Surprisingly often two separate authors, or sometimes even the same author, have used the same name to refer to two different taxa. This is particularly common with generic names when authors are trying to make them descriptive of the taxa concerned, but also occurs at family level and at species level. For example, there are probably many animals with obviously long tails or tail-like structures but only a limited number of obvious ways of expressing this in Latin or Greek, e.g. *longicauda* (*-is*) (*-um*). When names referring to two separate taxa of the same nomenclatural level and in the domain of the same Code, are spelt the same, the two names are called **homonyms**.

When it comes to species group names, two different types of homonymy can be distinguished, **primary** and **secondary homonyms**. Primary homonyms involve identical names for different taxa that were originally published in combination with the same generic name, even if one of these taxa is subsequently transferred to a different genus. Secondary homonyms are created when two species with the same specific name but originally described in different genera are subsequently brought together in combination with the same generic name. In both cases, the junior name is invalid and a new **replacement name** has to be proposed.

6.4.4 *The type concept*

Another important feature of the Codes, again with the exception of that governing viruses, is the use of the **type concept**, which means that material on which an original description is based has a special (type) status and will form the ultimate basis for any future determinations of the identity of a taxon.

Central to the purpose of nomenclature is that species and higher taxa once described should be recognizable by others in the future. Unfortunately, taxonomists are fallible, their descriptions may be incomplete or even inaccurate, and most importantly, they do not possess prescience and so their descriptions are unlikely to include mention of details that would be needed to separate their taxa from ones which might be discovered in the future. The idea arose quite early on therefore, that to enable future workers to identify previously described taxa unambiguously, it would be advantageous if they could have access to authentic reference material. Further, as any two taxonomists might, and quite frequently do, disagree about the exact limits of a taxon, the only material that can be guaranteed to correspond to the concept of a taxon in the sense of its original describer is the material that formed the basis of the original description. The material on which the description of a new species or subspecies is based is referred to as **type material** and has a fundamental role in taxonomy in that it fixes the meaning of a specific name.

With bacteria (not to mention several other groups) there is the additional problem that the diagnostic features of taxa are often not visible morphological characters but rather a collection of biochemical properties. Thus for these microorganisms it would be virtually pointless simply to preserve type specimens on a microscope slide. Instead, for bacteria the type concept is based on living cultures of the taxa concerned and not on dead preserved material. These cultures are termed **type cultures** and are deposited in a number of specialized collections around the world where they are maintained for as long as possible. As with more standard sorts of type material, such cultures may occasionally be lost (i.e. die out), and in these cases **neotype** cultures (neocultures) based on newly collected material, which appear on all known grounds to agree with the original cultures, can be started and maintained.

Just as species and subspecies are defined by their type material, so too are higher level taxa fixed by their types. *Effectively*, each higher taxon, e.g. genus, tribe, subfamily, family, etc., is typified by a single taxon of lower rank. Thus for example, the concept of a genus is fixed by its **type species**, and that of a tribe by its **type genus**. However, a small anomaly arises in the case of the ICBN in which the type of a supraspecific taxon is actually defined as the **type specimen** of one of its included species level taxa rather than the taxon itself. In practice this has little effect other than in the wording of articles.

In addition to allowing parts of an animal to be designated as the type specimen, for example, a bird could in theory be described from a feather or a beetle from a wing case, the ICZN also allows descriptions to be based on illustrations or photographs but in these cases the type is deemed to be the specimen illustrated and not the illustration itself. Thus if the Loch Ness 'monster', which was described formally as *Nessiteras rhombopteryx* for the first time in 1975 by Scott and Rines from a supposed underwater photograph, were to turn out to be genuine and not just a hoax, then the holotype would in fact be the specimen captured in the exposure. In this instance, one of the stated purposes of the description was to make protective legislation possible.

Recently, a conservation-minded action by four ornithologists has caused a considerable debate about the necessity of preserving a type specimen for certain new taxa if such an act might endanger the taxon concerned. In this case, Smith *et al.* (1991) described a new species of shrike, *Laniarius liberatus*, from Somalia on the basis of a single live caught individual which they subsequently released rather than preserving whole. Traditionally, type specimens of birds have consisted of a preserved skin and skull occasionally with additional skeletal remains and only rarely soft tissues. However, on the basis that the bird fauna of Somalia is relatively well known, Smith *et al.* concluded that the new species they had just caught might be extremely rare and therefore to kill it might have substantially reduced the survival chances of this species. Although the likelihood of these conclusions being true has been called into some question (see, for example, Peterson and Lanyon, 1992), this particular description raised even more interest because Smith *et al.* preserved blood samples and used DNA sequence analysis as an adjunct to their description giving the impression that the species was described on the basis of its DNA (see Hughes, 1992). Although the original diagnosis of *L. liberatus* was in fact based entirely on traditional morphocharacters, the combination of the lack of a traditional type specimen together with the prominence of DNA data in confirming this species' validity (for example, through ruling out the possibility that it was simply a hybrid of other better known *Laniarius* species) made this a widely discussed case. Many taxonomists worry that failure to preserve a standard type specimen means that many morphological characters that may be useful in future systematic research on the species will not be available. In this case, however, such arguments ought to be weighed against the undoubtedly honest concern of the original authors for the survival of the species.

6.5   Miscellaneous group-related factors

For historic reasons the Codes of nomenclature dealing with different kingdoms have been the responsibilities of different bodies (Commissions) and have been deemed to be independent of one another. Therefore, while many of the principles involved are similar or identical between the various Codes, there are also a number of differences both due to the biological peculiarities of the organisms concerned and due to the personal views held by the various people involved in the development of each Code.

6.5.1 *Animals and animal-like Protista*

Unlike botanical nomenclature, the provisions of the ICZN specifically exclude the naming of hybrid taxa since these are relatively uncommon among the animals whilst they are common among many groups of plants.

Due to the small size of most Protozoa and the frequent difficulty this presents in observing taxonomic features in preserved specimens, a single specimen if it were selected as a holotype might be virtually valueless as a means of delimiting and defining a species. In these cases therefore, the ICZN makes the special provision that the type of a species of Protozoa may be a group of directly related individuals preserved on a microscope slide, a **type slide**. Such a collective type is referred to as an **hapanotype**.

6.5.2 *Plants and plant-like Protista*

One of the difficulties encountered especially in botanical nomenclature is the need to deal with hybrids. In general, plant species belonging to a single genus hybridize readily with the production of fertile offspring. In contrast, in animals hybridization is relatively rarer, especially in the wild, and frequently results in sterile offspring even in crosses between closely related species.

6.5.3 *Fungi*

Despite now being nearly universally recognized as forming a separate kingdom of organisms from the Plantae, the systematics of the Fungi has traditionally being regarded as part of botany and it is still covered by the same Code (but see also Sigler and Hawksworth, 1986) although due to certain biological differences the Code makes a number of special provisions, particularly concerning the 'fungi imperfecti'.

6.5.4 *Lichens*

Lichens pose a particular problem for taxonomists in that they are the result of a symbiotic association between a species of fungus and an alga. Fortunately, although there are only a few species of alga ever found in lichen associations, virtually every type of lichen is formed by its own individual species of fungus and any lichen fungus is only found in association with a single species of alga, i.e. one lichen is effectively equivalent to one fungus. For nomenclatural purposes, therefore, lichens are covered by the International Code of Botanical Nomenclature and the names given to them are deemed to apply to their fungal components. Of course things are seldom completely straightforward and it is worth noting that a handful of lichen-forming fungi do form associations with two different algae and the resulting lichens have quite different forms. In these cases only detailed investigation can show that the two lichens are actually a single 'species'.

An interesting analogy to the situation in lichens can be seen in some freshwater sponges in which intimate associations between two good species have at times given rise to composite organisms that have been described as separate species (Ricciardi and Reiswig, 1992). Of course, such composite species have to be regarded as (partial) synonyms of the two pure species involved.

6.5.5 'Blue-green algae' (Cyanophyta versus Cyanobacteria)

This group of photosynthetic prokaryotes has been the source of many nomenclatural difficulties. As these organisms are photosynthetic and commonly form colonies with distinct morphologies, they were for many years covered by the ICBN. Yet while their nomenclature is substantially embedded in the botanical literature with more than 2000 'described' taxa, in evolutionary terms there can be no doubt that their closest allies are the rest of the Prokaryota (Monera), namely the bacteria, and so they are also covered by the rules of the ICNB. The problem therefore arises as to which nomenclatural Code they should be incorporated under (Friedmann and Borowitzka, 1982). Logically, they should become fully assimilated into bacteriological nomenclature and their historical botanical ties should be relinquished.

On a second front, consideration of 'blue-green algae' taxonomy presents a nightmarish scenario. The group is without doubt one of the most plastic known in terms of the form of their colonies which are determined to a great extent by the prevailing ecological conditions. This means that any one taxon can often develop into a range of markedly different vegetative forms according to field conditions. This flexibility has led some workers to develop substantially different classifications, in one instance reducing the number of species to just

62. What is clear is that the group is in need of a major systematic study including more experimental work on the influence of ecological factors and the incorporation of modern ultrastructural and molecular genetic information as well as standard morphology.

6.5.6 *Bacteria and other prokaryotes*

With bacteria, the marriage of phylogeny, systematics and nomenclature poses real problems. First, it will be obvious that these organisms do not possess a large number of external morphocharacters and their identification relies on a wide array of staining, fermentation, digestion and drug resistance tests and even on DNA base ratios (Buchanan and Gibbons, 1974; see section 11.4). Second, even the definition of a species is not as easy as for most plants and animals since for the most part bacteria do not form interbreeding populations but rather form clonal colonies with only limited genetic exchange between gene lines (**genospecies**).

Due to the fact that bacteria have only a few obvious external features and due to the failure of many earlier bacteriologists to make an effort to preserve living type cultures, bacterial systematics made a revolutionary clean sweep with the 1976 revision of its Code (Lapage *et al.*, 1975). A new start date for bacterial nomenclature was set as 1st January 1980 (corresponding to the publication of Skerman *et al.*, *Approved Lists of Bacterial Names*, 1980).

6.5.7 *Viruses*

Viruses also pose certain problems for the taxonomist (Matthews, 1979). In many ways their comparatively recent discovery and the difficulty of making direct morphological observations may actually help to avoid some of the problems that bedevil bacteriological taxonomy but obtaining universal agreement from workers on all the different groups of viruses may well prove very difficult. For this reason, viral nomenclature is still a patchwork with different traditions being applied to different groups, a situation not aided by the fact that many virus workers are more or less pure molecular biologists with no interest in virus taxonomy.

Viruses are only classified into two higher categories, the genus and the family (Table 6.2). Because their taxonomy is a relatively recent occupation it should come as no surprise that many viruses are not yet assigned to genera or even to families. Broadly, viruses are divided into those which attack bacteria (the phages), vertebrates, invertebrates, plants and fungi although there are a few groups (families) that have members which collectively attack vertebrates and invertebrates or invertebrates and plants or occasionally all three. Phages are not assigned to genera and in fact many are not yet assigned to families. Plant viruses are for the most part assigned to families but some are known by vernacular names or standard abbreviations thereof (e.g. tobacco mosaic virus, TMV) whereas others are given generic and species names like higher organisms. Animal viruses again are like plant viruses with a mixture of vernacular and scientific names.

Virus families are not included in a hierarchy of higher level taxa for two good reasons, first, they are usually insufficiently well known and second, and perhaps more importantly, it is likely that viruses have evolved on many separate occasions from their host organism's genomes and therefore to place families in orders and classes would tend to imply some degree of relatedness which might not be there at all.

6.5.8 *Organisms showing extreme polymorphisms*

In a number of groups, biological factors have lead to particular nomenclatural problems because within the life cycle of an individual or between different generations of a species there can be such extreme differences in form that without biological studies earlier taxonomists would have been unable to associate one form of the species with another. In many instances, for example the larval and adult stages of butterflies and moths, this did not matter because taxonomy of the group was traditionally based on only one developmental stage, in this example the adult or imago. However, in some groups there was a need from the beginning to name members of both forms. Notably among these are those Cnidaria (Coelenterata) in which the life cycle includes a polyp stage and a medusa (jellyfish) stage. Not surprisingly in these, two independent nomenclatures were developed for each of the stages, and it is only comparatively recently that extensive rearing has enabled the medusa and polyp stages of the majority of temperate species to be associated with one another. The ICZN does not lay any restrictions on which names should be valid, and thus the correct names are determined by priority of publication in the absence of the use of plenary powers to conserve one name. This is in marked contrast to the next example.

In the mosses (Bryophyta) a considerable problem emerges due to the alteration of generations between the haploid gametophyte stage and the diploid sporophyte stage. Whereas for other land plants the taxonomy has drawn almost entirely from the characters of one stage (the diploid), that of mosses has utilised information from both to a considerable degree. The situation appears to be confounded by the fact that there is little congruence between classifications based on the two separate stages. Part of the problem here may stem from the dominance of phenetic methodology rather than of cladistics in constructing moss classifications (Rohrer, 1988).

Another problem group is the vast number of **pleomorphic** fungi. These are fungi which have two separate types of spore-producing structure, one sexual (i.e. as a result of meiosis) and one asexual (i.e. as a result of mitosis). The sexual spores are produced by **teleomorphic** morphs whereas asexual spores are produced by **anamorphic** morphs. Unfortunately, for many taxa only one or the other morph was known at the time of publication therefore giving two potentially conflicting classifications. To complicate matters further, the classification that resulted from the common early practice of simply considering the final arrangement of anaspores is not particularly congruent with that obtained from study of teleomorph structure. Nowadays it is realized that a better indication of systematic position of anamorphic fungi is given by study of the development of its spores rather than their final arrangement, although this does not help the earlier anamorph classifications. To overcome this, the ICBN specifies that valid names can only be applied to taxa for which the type specimen is a teleomorph. When it is necessary to describe or refer to a taxon known only from its anamorph, the Code considers these names to be **form-taxa**. Prior publication of form-taxa names does not give them priority over the first publication of a valid name for the teleomorph irrespective of the dates of publication.

6.6 Names of higher groups

In all Codes except for the ICNV, some provision is made to specify how the names of suprageneric taxa should be formed. Specifically, names of taxa of particular levels should terminate in certain specified endings thus enabling a reader to know to which hierarchical category reference is being made. The Codes have, however, been individualistic in this respect, thus while the ICBN specifies or recommends endings for all levels from division to subtribe, the ICZN only deals with the categories from superfamily to tribe (although there is also a widely used termination for subtribe) (see Table 6.2). Even the names of the categories (ranks) of higher taxa that are recognized by the Codes differ, for example, below kingdom (often referred to as Regnum in botanical nomenclature) in the plant and bacterial nomenclatures comes division, whereas the equivalent in zoology is phylum. Proposals such as those of Rasnitsyn (1982) to regulate the forms (e.g. derivation and suffixes) of names of higher taxa have generally received scant support because to impose such regulations would result in an extremely large number of changes.

6.7 Starting dates for nomenclature

In order to achieve stability in nomenclature while still holding to the general principle of priority it has been decided for each group of

Table 6.2 Standard suffixes for family group names of animals, plants, fungi, bacteria and viruses. Suffixes in bold-type are mandatory except where special provisions have been made, others are recommendations or effectively standard through usage.

Taxonomic *level	Suffixes					
	Animals	Higher plants	Algae	Fungi	Prokaryotes	Viruses
Division	NA	*-phyta*	*-phyta*	*-mycota*	—	NA
Phylum*	—	NA	NA	NA	—	NA
Subdivision	—	*-phytina*	*-phytina*	*-mycotina*	—	NA
Class	—	*-opsida*	*-phyceae*	*-mycetes*	—	NA
Subclass	—	*-idae*	*-phycidae*	*-mycetidae*	—	NA
Order†	—	**-ales**	**-ales**	**-ales**	**-ales**	NA
Suborder†	—	**-ineae**	**-ineae**	**-ineae**	**-ineae**	NA
Superfamily	*-oidea*	NA	NA	NA	NA	NA
Family	**-idae**	**-aceae**	**-aceae**	**-aceae**	**-aceae**	**-viridae**
Subfamily	**-inae**	**-oideae**	**-oideae**	**-oideae**	**-oideae**	NA
Tribe	*-ini*	**-eae**	**-eae**	**-eae**	**-eae**	NA
Subtribe	*-ina*	**-inae**	**-inae**	**-inae**	**-inae**	NA
Genus	—	—	—	—	—	**-virus**

* Division and phylum are equivalent ranks in botanical and zoological nomenclature respectively.
† The endings for order and suborder names governed by the botanical code are only mandatory if they are based on the name of an included family.
NA, not applicable.

organisms that there needs to be a definite starting date for their nomenclature, otherwise people might keep on discovering older, effectively binomial names for organisms dating back into antiquity as more and more obscure literary sources are searched. However, there is no single date on which Linneaus's binomial system was applied to every group. Linneaus himself was predominantly interested in botany and it is not surprising therefore that his first exposition of the binomial system in 1753 dealt with plants and that the application of his system to animals did not appear until five years later.

Having an official start to the nomenclature of organisms means therefore that if a Latin binomen for a particular taxon was published before the relevant official start date, it would not be considered as valid by the Code, i.e. it would have no current nomenclatural status. The currently defined starting dates for nomenclature are different for different groups as shown in Table 6.3. For convenience, the Codes sometimes deem particular works to have been published on particular dates so that they either partake in or are excluded from considerations of priority, synonymy, etc.

6.8   Citation of authors

Because the same generic names or even whole binomens have sometimes been used by various authors to refer to different taxa (homonyms; see section 6.4.3), a scientific name or a generic name is not truly complete unless the name of the author who originally described the species is given afterwards. This **citation** avoids ambiguity about to which taxon reference is being made.

In the case of specific names the situation is complicated a little because the author of a species may originally have placed it in a different genus to that in which it is now placed. It is easy to see that such changes might be confusing to someone trying to track down an original description or other publication on that species. Therefore an additional convention is employed to distinguish between authors' names when they refer to the same combination as that published originally as opposed to when the combination results from the subsequent realignment of a species with a new genus in which case the author's name is given in parentheses.

The citation of authors in botanical and zoological nomenclature is essentially similar although in practice zoologists in citing authors

Table 6.3 Starting dates for nomenclature of different groups and their base publications

Taxonomic group	Starting data	Base publication
Multicellular animals and animal-like Protista (including fossils)	1st January 1758	10th edn Linnaeus's *Systema Naturae*
Metaphytan plants excluding plant-like Protista (Algae): (Spermatophyta), ferns (Pteridophyta), liverworts (Hepaticae) and sphagnum mosses (Sphagnaceae)	1st May 1753	1st edn Linnaeus's *Species Plantarum*
mosses (Musci)	1st January 1801	Hedwig's *Species Muscorum*
fossil plants*	31st December 1820	1st vol. Sternberg's *Flora der Vorwelt.*
Fungi (including lichens)*	1st May 1753	1st edn Linnaeus's *Species Plantarum.*
Algae (i.e. plant-like Protista) most groups	1st May 1753	1st edn Linnaeus's *Species Plantarum*
Desmidiaceae	1st January 1848	Ralfs *British Desmidieae*
Oedogoniaceae	1st January 1900	Hirn's *Monographie und Iconographie . . .*
Blue-green algae (Cyanobacteria)†		
Nostocaceae (Homocysteae)*	1st January 1892	Gomont's *Monographie des Oscillariées*
Nostocaceae (Heterocysteae)*	1st January 1886	Bornet and Flahault's *Révision des Nost . . .*
Bacteria and other Prokaryota	1st January 1980	Skerman *et al.*'s *Approved List . . .*
Viruses	No official starting date	

* In these groups the ICBN makes special provision to allow names published in other specified places either before or after these dates to be considered as valid or to be effective from these dates.
† These starting dates are those specified by the ICBN and not by the ICNB which also regulates the nomenclature of Cyanobacteria, treating them just like all other Prokaryota.

responsible for transferring a species to another genus commonly distinguish this author from the original one simply by inserting a colon between the citation and the species name; the ICZN in fact recommends that the original authors (in parentheses) be mentioned too. A second trivial difference is in the use of abbreviations of the names of authors, a process recommended by the ICBN but not generally practised in zoological nomenclature except for the names of a few famous early taxonomists, for example, Linnaeus is often simply referred to by 'L.'. The ICBN's recommendation would make more sense were some authoritative list of authors and their abbreviations to be published.

6.9 Publication

One of the functions performed by the Codes is to define what exactly makes a nomenclatural act (e.g. the description of a new taxon) a valid one, and hence enables identification of names which can be regarded as **validly** (effectively) published and which cannot. However, the term publication is subject to a wide range of possible definitions and the media that are available in which to publish are changing ever more rapidly with, for example, desk top publishing, electronic media, etc. Therefore, each of the Codes sets down in a precise way what exactly constitutes an act of publication as regards their particular coverage. In general, for modern descriptions, publication must involve the production of a minimum number of printed copies which must be made generally available by sale or gift. Nicolson (1980) provided an interesting key to what constituted and did not constitute valid publication of botanical names at that time.

The first publication of a scientific name (genus or species) must be accompanied by an adequate **original description** that should in principle enable its separation from previously described taxa. Sometimes, however, authors have failed to present an original description, or at best these may end up being published some time after the first published mention of the name. In the former case the name is referred to as a **nomen nudum** or **naked name**.

For names of extant organisms covered by the ICBN (i.e. plants, fungi and Cyanophyta/Cyanobacteria) there is an additional requirement that for valid publication the name must be accompanied by a description of diagnosis in Latin. For those concerned with this process, a useful guide book to botanical Latin has been written by Stearn (1973).

6.10   Type depositories

In order that it may be kept safely and be made accessible to any taxonomist needing to study it, type material should be deposited in one or more major national/international collection where the appropriate safeguards to curate and protect it exist while still permitting unrestricted access to bona fide researchers. Ideally, all type material of a given taxonomic group in a single country should be housed in one major institution. Unfortunately, this is often not the case and a researcher may need to visit many separate institutions.

In the case of cultivable bacteria, type material is in the form of living cultures which may need routine subculturing and definitely do when a researcher needs to study them. Thus when a new species of cultivable bacterium is described, there is a more limited, although still impressively large, number of collections where type material must be deposited (Lapage *et al.*, 1975). Because there is a potential for losing these living cultures there is perhaps an even greater need to deposit replicate type cultures in several separate institutions.

6.11   Good practice

Apart from the rules set down in the Codes of nomenclature, the Codes also contain a large number of **recommendations** for good practice covering such things as how to decide on what taxa to designate as a type or which specimens from a paratype series to choose as a lectotype, how to form new scientific names, etc. Whereas the rules are mandatory* the recommendations are not, although if a taxonomic manuscript contravenes them there is a good chance that it will be rejected from 'quality' refereed journals until it has been appropriately modified.

In addition to the recommendations presented in the various Codes, any author of a taxonomic work should always bear in mind the needs of users who may not be so expert in the group. Some more general aspects of good practice include the routine provision of keys and good

---

* Mandatory here is relative in that if a taxonomist does not follow the rules and proceeds with valid publication (by whatever means), then the actions taken, such as publication of new names or synonymies, still have to be taken into account by subsequent workers.

illustrations, the extensive citation of the field so that the inexperienced can gain access to relevant literature, the clear citation of all material examined, etc. Many of these points are discussed and illustrated by Wiley (1981).

Even given these recommendations, one important issue still remains, namely how taxonomists should decide if they are dealing with an undescribed species. In practice, much relies on the experience and judgement of the individual taxonomist which will be influenced in no small way by their knowledge of the usual degree of interspecific variation within the group concerned. In many groups, for example, coloration can show considerable interspecific variation and is often polymorphic. In these cases it is obvious that coloration should not be used by itself as a basis for describing a new species. However, in other groups coloration may be very consistent within species and it could even provide the only readily accessible means of distinguishing between closely related taxa. Thus, there can be no hard and fast rules about what are sufficient criteria for describing material as a new species, and any description, particularly one based solely on preserved material, might be proved unsound in the light of new biological data or genetic data or even the discovery of new intermediate specimens.

Basing descriptions of new taxa on single specimens is always risky since there is usually the possibility that it is an aberrant individual or a hybrid. The more identical specimens there are the more likely they are to represent a genuine new taxon. Nevertheless, many taxonomists do publish descriptions based on singletons.

Another consideration concerns what to do if a taxonomist has discovered a previously undescribed species. While new species of well-known groups such as the birds turn up only rarely, for many groups museum collections may abound in specimens of undescribed taxa. In the latter case, a taxonomist could spend a lifetime just describing new species that in the end would benefit science very little, some would even say that increasing the numbers of descriptions in large but poorly known groups is a disservice to science, cluttering the literature, etc. Four factors in particular should be considered in this respect: the intrinsic interest of the group (for example, new bird or mammal species engender a wide interest), the potential importance of the new taxon for other areas of biology (e.g. is it a pest or beneficial species or the subject of some biological observation?), does the new species provide important new evidence about the phylogeny, biogeography, etc. of the group, and is the new taxon going to be included in a revision?

6.12   Major taxonomic publications

While taxonomic matters are dealt with by many journals with varying frequencies, there are a few serial publications that have special or important roles in taxonomy. The Codes of botanical, bacterial and viral nomenclature are published as parts of series or in a limited number of journals (see section 6.6); for example, the ICBN and ICNCP are published as parts of the irregular serial publication, *Regnum Vegetabile*. The proposals and rulings of the Commissions of Botanical and Zoological Nomenclature on the conservation and rejection of names or the sanctioning of works are published in *Taxon* and the *Bulletin of Zoological Nomenclature*, respectively. All new names of bacteria must be announced in the *International Journal of Systematic Bacteriology*. Other journals which carry a high percentage of articles relevant to taxonomic methodology are *Annual Review of Ecology and Systematics*, *Cladistics*, *Evolution*, *Systematic Botany* and *Systematic Zoology*.

CHAPTER SEVEN

CYTOTAXONOMY

Many biosystematic studies over the past 40 years have included cytological observations, but all too frequently the authors have not derived the maximum amount of information from their preparations.

Jackson (1984)

7.1 Introduction

Chromosomes have been recognized as taxonomic characters for a very long time but, as indicated by Jackson (1984), a large percentage of cytological investigations have unfortunately failed to provide taxonomists with as much useful information as they could have, even though the workers may have put considerable effort into their studies. Thus if they were principally interested in cytogenetics they would frequently fail to record information pertinent to taxonomy. Conversely, if the work was conducted by taxonomists, their frequent lack of cytogenetic training means that they often missed many potentially informative pieces of data.

7.2 Karyotypes

In eukaryotic organisms, the nuclear genome is organized into chromosomes, the number of which varies between species from a haploid number (n) of 1, for example, in the ant, *Myrmecia pilosula*, and the nematode worm, *Proascaris*, to more than 200 in some butterflies and as many as 250 in some ferns (Imai and Taylor, 1989; Sivarajan, 1991). Whereas genetically controlled variation in other characters is ultimately the result of differences in nucleotide sequences *within* expressed genes, major physical rearrangements of the genome within chromosomes will usually have no apparent effect on the organism's normal development, physiology or behaviour as long as the relative numbers of functional genes remain undisturbed and they are not separated from their control regions. However, chromosomal rearrangements can have a profound influence on the fertility of offspring resulting from crosses between individuals of different **karyotypes**, and therefore they have considerable evolutionary significance (Jackson, 1976, 1982; Maynard Smith, 1989). Furthermore, karyotypic information effectively forms an independent data set for phylogenetic analysis and has probably been most useful in the investigation of groups of closely related and morphologically similar organisms.

The interpretation, characterization and identification of a cell's complete chromosome set is referred to as karyotyping and is the first stage in the process of using chromosomal characters for systematics. For many organisms karyotyping is relatively easy if an appropriate protocol is followed, however, in some groups the presence of large numbers of small chromosomes or of generally very small cells can make this a more complicated process. For example, most birds have a small number of normal sized chromosomes but also have a large number of microchromosomes which may be extremely difficult to count, let alone identify.

Probably because of the widespread occurrence of reduced fecundity in crosses between individuals of different karyotype (hybrid disadvantage), karyotypes within interbreeding populations of a species are usually remarkably constant although a few marked exceptions occur. However, it is apparent that at least in small isolated populations, modified karyotypes may become fixed with apparent ease (see below), and therefore karyotypes can provide much useful information about the evolution of a group, the more so as geographic isolation is probably the most important mode of speciation.

Karyotypic variation may involve several types of change. Whole chromosome sets may be gained through **polyploidy** events (which themselves can take several forms). Rearrangements of chromosome sections can occur (e.g. translocations and inversions; see section 7.5) or individual chromosomes may be lost or gained either by errors in their migration to daughter cells during meiosis or, probably more important, the fusion of one chromosome with another (combined with subsequent loss of the extra **centromere**) or, less commonly, chromosome fragmentation.

The degree of variation in chromosome number between closely related species is subject to great variation. In some groups even closely related species can be widely dissimilar as epitomized by the muntjack deer with the Chinese subspecies having a haploid number (n) of 23, while the Indian subspecies has only 3 plus one sex chromosome (Benirschke and Kumamoto, 1991). Even more extreme is the variation encountered in the ant genus *Myrmecia* in which n ranges from 1 to 42 (Imai and Taylor, 1989). In marked contrast to these examples, considerable consistency in chromosome number can sometimes be observed throughout major groups of organisms. Thus, for example, in the Lepidoptera (butterflies and moths) and the Trichoptera (caddisflies), which diverged from one another at least 65 million years ago, during the late Cretaceous, the commonest chromosome number is approximately the same ($n = 30$ or 31). Even when chromosome number differs significantly from these values, the total genomic DNA content is actually stable, indicating that polyploidy has not been a significant factor in these insects. A similar consistency in number is also shown by the bony fish in which $n = 24$ is by far the most common value although the overall range goes from $n = 11$ to $n = 60$ (Hinegardner and Rosen, 1972).

A particular problem in this respect can crop up when **autopolyploidy** of various degrees occurs, as in quite a number of plants. In those cases in which morphological characters can also be found to distinguish between plants of different ploidy level it is standard to regard the different karyotypes as different species. However, in some cases no morphological difference can be found between individuals with widely different chromosome numbers (e.g. the bedstraw, *Galium aparine* includes individuals with $2n = 22$, 44, 66 and 88). In these cases they are sometimes retained in the same species even though the consequences for interbreeding are often not fully understood.

Chromosome number alone is typically a poor indicator of phylogeny because changes can originate through multiple roots, e.g. it will be reduced by one no matter which of a set of chromosomes is lost or which of probably many possible pairwise combinations become fused. Thus, in order to use chromosome number more meaningfully, it is necessary to be able to identify individual chromosomes.

Identification of homologous chromosomes in specimens from the same or closely related species can be difficult. Fortunately, in many species, some or all of the chromosomes may have distinctive and consistent morphological and/or chemical features that can enable recognition between individuals. Morphological differences in form include chromosome size and the locations of their centromeres. In the salivary glands of some flies, particularly large **polytene** chromosomes are found which have a natural banding pattern that can enable recognition not only of individual chromosomes but also of small segments of chromosomes. In many other species, differences in the densities of **chromatin** and **heterochromatin** along chromosomes and differences in the relative amounts of A–T (adenine–thymine) rich or C–G (cytosine–guanine) rich DNA along chromosome arms can be revealed by appropriate staining procedures, again permitting the recognition of particular chromosomes or parts of chromosomes.

A cursory consideration of the karyotypes of a few different organisms will soon show that while some are dominated by **metacentrics** others have high proportions of **acrocentrics** and/or **telocentrics** (Figure 7.1). Such differences are often apparent even between quite closely related taxa such as species within a genus or genera within a tribe. Unfortunately, changes in chromosome number or, in the case of plants, ploidy level may make it effectively impossible to score the character state of homologous chromosomes in the different taxa and therefore if this variation is to be used for taxonomic purposes it is frequently necessary to obtain an index of chromosome arm asymmetry. An appropriate measure is the **intra-chromosomal asymmetry index**, A_1 (Zarco, 1986), where

$$A_1 = 1 - \left(\frac{\sum_{i=1}^{n}(b_i/B_i)}{n} \right)$$

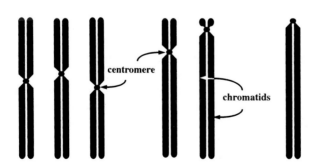

Metacentrics   Acrocentrics   Telocentric

Figure 7.1 Diagrams of three different chromosome morphologies.

where b_i and B_i are the lengths of the shortest and longest arms of the n chromosomes in the karyotype. This index is framed so as to be close to zero if all the chromosomes are metacentric and near one if all are telocentric.

7.3   Chromosome banding

DNA is inevitably non-homogeneous along the arms of a chromosome. During mitosis and meiosis, local variations in the degree of condensation of the chromatin to form heterochromatin may be apparent in metaphase preparations. Other differences such as the ratio of adenine + thymine to cytosine + guanine, and the concentration, composition and lengths of repetitive sequences may be visualized by more specific staining procedures.

The banding patterns of the giant polytene chromosomes from the salivary glands and certain other tissues of larval Diptera (true flies) have proved particularly useful in both distinguishing cryptic species and even for reconstructing their phylogenies (Dunbar, 1966). Such work is of considerable importance as one of the groups most extensively studied to date is the blackfly family, Simulidae, which includes many species that are vectors of serious diseases, especially in the tropics. Cytotaxonomy of these flies has been greatly facilitated by their polytene chromosome banding patterns. Within the Simulidae, chromosome number is constant with a haploid number of 3. The *Eusimulion aureum* group are atypical in having $n = 2$, but examination of their banding patterns shows that in this species group the two smaller simuliid chromosomes have become fused.

Despite their immense usefulness in the cytotaxonomy of true flies, polytene chromosomes occur only rarely in other organisms (for example, in the springtails (Collembola) and in dinoflagellate Protista) and therefore recognition of chromosomes or parts thereof in other groups depends largely on the use of more sophisticated staining procedures.

Banding patterns can be revealed in many non-polytene chromosomes by appropriate staining protocols such as **C-**, **G-** and **Q-banding**, brief accounts of which are given in the glossary. These have had a patchy use in cytotaxonomic studies to date. Nevertheless, where they have been employed they have often helped considerably in identifying chromosomal homologies, inversions, translocations, etc., and have, like other characters, been employed to produce cladograms.

The ability of chromosome banding to distinguish between otherwise similar chromosomes has also revealed some interesting facets of the organization of mitosis and meiosis in polyploid hybrids. Cell division turns out to be even more sophisticated and organized than might at first be imagined, as in hybrid cells the chromosomal sets from each parent are not simply randomly assorted during division but at metaphase the chromosomes from one parent tend to form a ring around those from the other parent (Linde-Laursen and van Bothmer, 1988).

7.4 Chiasma frequency

Chiasmata or crossovers are evident in meiotic cell divisions in **diplotene**, when they provide visual evidence of genetic exchange between sister chromosomes. Chiasma frequency is itself under genetic control and the number of crossovers observable in meiotic cell preparations tend to be relatively constant within subspecies or species and often differ between species or subspecies. Their frequency can therefore be used as a taxonomic character.

Chiasma frequency can also be used as an estimator of the homology between the two sets of chromosomes occurring in hybrid individuals by comparing them with the numbers found in each of the parental species (Jackson, 1984). If the two sets of chromosomes have a high level of homology, then no depression in the number of chiasmata would be expected. Generally, it has been found that interspecific hybrids, and even hybrids between different strains of a single species, show a reduced frequency of crossing over.

7.5 Inversions, translocations and their significance

Inversions and **translocations** are two forms of chromosomal rearrangement with considerable implications for the genetic compatibility of sexually reproducing organisms. In an inversion a segment of a chromosome becomes reinserted in the same chromosome but the opposite way around (Figure 7.2) while in a translocation, a chromosomal segment is removed from one place and reinserted somewhere else in the genome, either in the same or in some other chromosome. Further, inversions can either involve only one chromosomal arm or they can incorporate the centromere, so-called **paracentric** and **pericentric** inversions, respectively. Inversions and other chromosomal rearrangements have perhaps been most thoroughly investigated in the true flies (Diptera) because of their naturally banded giant polytene chromosomes which make them particularly easy to recognize and allow accurate definition of the chromosomal segments involved.

The significance of these chromosomal rearrangements to taxonomy is two-fold. Firstly they are characters that can be recognized, for example, in polytene chromosomes or in other material with suitable histological staining procedures, and then used in phylogenetic analyses. Secondly, pairing during meiosis between a normal chromosome and one with a sufficiently large translocation or inversion can potentially have a profound effect on the fertility of hybrid organisms. During meiosis pairing takes place between the homologous chromosomal arms of maternally and paternally derived haploid chromosome sets and chromosome segments are exchanged between these pairs. Under normal circumstances each of the resulting recombined chromosome arms contains the same genetic complement but with the genes from the two parental sets having been reshuffled between homologous chromatids. However, when a chromosome in one of the haploid sets possesses an inversion with respect to its homologue,

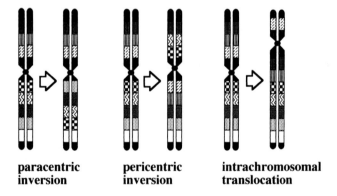

paracentric   pericentric   intrachromosomal
inversion   inversion   translocation

Figure 7.2 Diagrams showing the difference between two forms of chromosomal inversion and a translocation.

pairing and resulting exchange of chromosomal segments may result in one of the recombined chromatids missing some genes and the other having two sets. Any zygote formed involving gametes that either lacked or had duplicated sets of genes would be unlikely to survive because of their unbalanced genomes. In the case of pericentric inversions even greater problems occur for hybrids that were heterozygous for the inversion as one of the chromatids could end up with two centromeres while the other would have none, with the result that chromosome segregation would be disrupted.

Whether chromosomal rearrangements *per se* account for many instances of the reduced fecundity often observed in interspecific hybrids is still debated after many years. Certainly there are plenty of examples in which the chromosomes of the members of species pairs differ quite markedly in terms of inversions, translocations or chromosome fusions, yet hybrids between them show little or no reduction in fertility, for example, various *Drosophila* or grasshoppers. Nevertheless, White (1982) argues that these may be exceptions and that there are also plenty of counter examples in which small chromosomal rearrangements have dramatic effects. In any case, any chromosomal rearrangement after its first occurrence must exist as a polymorphism within a population, and therefore if this rearrangement were to be disadvantageous in meiotic pairing with non-rearranged chromosomes then it would be selected against from the beginning and would never spread through the population. If this were so, then only chromosomal rearrangements that were not disadvantageous would ever persist or become fixed in a population. However, several arguments and mathematical models suggest that the above argument is too simplistic. In short, rearrangements that could be disadvan-tageous in the heterozygous form (i.e. as in interspecific hybrids) could become fixed if either populations were effectively subdivided into small units (**demes**) and/or there is considerable inbreeding and/or individuals homozygous for the new arrangement are at an advantage.

The importance of chromosomal rearrangements for phylogenetic analysis stems from their supposed uniqueness. In this respect it has been argued that since a chromosome could split and rearrange at any of many possible positions, it is extremely unlikely that a given chromosome inversion or translocation could undergo a reversal back to exactly its same original state. It might also be argued that it is unlikely that identical chromosomal rearrangements could arise more than once in the evolution of a species. However, while this may be true at the molecular level, there is still the practical problem of being able to recognize the difference between two different chromosomal rearrangements that involve similar but non-identical transformations. Nevertheless, the proposed uniqueness of any particular inversion appears to be largely supported by other evidence although there may be occasional exceptions. As explained by Farris (1978), unique acquisition of an inversion does not rule out reversal because while the back inversion may be extremely unlikely, a population could be polymorphic for the inversion and from this situation the new inversion could either become fixed or could be selected against and ultimately eliminated. Phylogenetic analysis of chromosomal rearrangements might therefore be better conducted under the polymorphism parsimony criterion rather than with the assumption that chromosomal rearrangements are irreversible (see chapter 2 and Figure 2.10).

7.6   *In situ* hybridization

The *in situ* hybridization procedure relies on the fact that under appropriate circumstances a length of single-stranded DNA will bind to an homologous stretch of DNA in a cell nucleus or chromosome and so make it possible to locate whereabouts in an organism's chromosome set a particular gene sequence occurs. In practice, an artificially synthesized probe sequence of DNA is labelled either radioactively or nowadays often with a highly fluorescent chemical group. It is then allowed to hybridize with chromosomal DNA in a cytological preparation on a microscope slide and after washing off the excess unbound probe the cytological preparation can be stained and either autoradiographed or examined under UV to reveal the location of the bound probe among the chromosomes.

   In general, this technique is usually used to locate genes that are represented within the genome by many copies, for example ribosomal RNA genes, tRNA genes, certain repeated sequences, etc. because the level of radioactivity or fluorescence available from a single probe molecule is low and so will not give a clear signal above background. Nevertheless, some studies with radio-labelled probes have shown that by counting silver grains over the chromosomes in many spreads, the location of a gene represented by a single copy (or perhaps a few copies) can be deduced.

   A useful taxonomic role of *in situ* hybridization has been in the verification of hypothesized chromosomal homologies based on general morphology and banding procedures as, for example, is illustrated by the study of Steinemann *et al.* (1984) with chromosomes of the *Drosophila obscura* group. This sort of confirmatory role can greatly increase confidence in phylogenies based on putative chromosomal rearrangements. Another growing use of *in situ* hybridization is in genome painting, a technique which is gaining in use for the detection of taxa that have resulted from interspecific hybridization events such as appears to be particularly common among the higher plants (Schwarzacher *et al.*, 1989; Leitch *et al.*, 1991; Bennett *et al.*, 1992). In this case use is made of naturally extracted and subsequently labelled DNA from one species which is suspected of being one of the parents of the supposed hybrid. The DNA, after purification, is cleaved into fairly short fragments that are then labelled at one end with a fluorescent marker. Subsequent hybridization of these labelled fragments with cytological chromosome preparations may show more or less even binding indicating generally spread sequence homology, or it may show that the preparation's chromosomes fall into two distinct groups which could indicate that the species under investigation is indeed a hybrid with one part of its genome derived from the species providing the DNA probe. Further confirmation can be obtained by investigating which chromosomes appear to have been derived from the other parent, and it is even possible to use double labelling procedures with DNA from one parent labelled with a fluorochrome of one colour while a differently coloured marker is used for DNA from the other parent. Further, such studies have helped reveal that in hybrids, chromosomes do not intermingle at random during cell division but rather that those of one parent tend to surround those of the other during metaphase.

CHAPTER EIGHT

CHEMOTAXONOMY AND RELATED TOPICS

Taxonomists have great experience in distinguishing what characters may
or may not be significant in morphology and anatomy, but significance is
not so obvious to them in chemistry.

Smith (1976)

8.1 Origins of chemotaxonomy

In the sense employed here, chemotaxonomy is taken to refer to the
taxonomic use of information about principally small molecules that
are generally produced by the actions of enzymes, and it is not taken
to include information about proteins and nucleic acids.

All organisms synthesize a plethora of low to medium molecular
weight chemicals apart from proteins and nucleic acids, which
collectively serve many different functions ranging from fundamental
roles in essential metabolic pathways and inter- and intracellular
communication, to perhaps more flexible areas of coloration and
communication. As the production of these compounds is largely
determined by the enzymes encoded by the organism's genome, such
compounds are essentially similar to morphological characters in that
the presence of the compound implies something about expression of
one or more functional genes. There has been considerable debate as
to whether such chemical characters are intrinsically better or worse
than traditional morphocharacters for inferring relationships.

8.2 Classes of compounds and their biological significance

That different species contain substances with different properties is
deeply embedded in folklore even if understanding the chemical bases
of these differences is only a comparatively recent development. The
Bushmen of the Kalahari along with many other tribal peoples use

smell and taste as diagnostic characters for many of the plants they encounter and use. Such features are no doubt particularly helpful for distinguishing leafless twigs or tubers; however, with a few exceptions such as the identification of some lichens, 'western' botanists almost totally ignore such ancillary characters for the identification of the vast majority of plants they encounter.

8.2.1 *Sex pheromones*

Many organisms use chemical signals to attract mates, to signal to potential mates their true specific identify and to indicate sexual receptiveness or otherwise. It seems likely therefore that these sex pheromones must be tailored in order to avoid wasting time and effort in attracting individuals of different closely related species and therefore will carry particularly valuable information relevant to species identity. Consider, for example, a temperate woodland with perhaps 1000 species of moths seeking mates at the height of the season. The potential for confusion here is obvious and in the tropics, where diversity is generally greater, the situation is even more complex. Therefore it is likely that there is very strong selection pressure within species for the conservation of sex pheromones and between species for their differentiation. After all, an individual giving out the wrong signal is likely to be fundamentally disadvantaged especially if interspecific mating and consequent production of infertile offspring were to result.

8.2.2 *Lipids and hydrocarbons*

Cuticular hydrocarbons have been a major source of taxonomic information for many insects, especially social groups such as ants and termites. Their advantage lies in their ease of extraction, separation and identification using gas chromatography and mass spectroscopy. The results of such analyses are traces showing the relative abundances of the various constituents of what are typically complex mixtures of compounds. Although it is frequently the case that the chemical identities of many of the constituents are unknown, the data obtained generally show considerable consistency within species and often vary markedly between even closely related species. In some groups, hydrocarbon traces have provided powerful evidence of the existence of cryptic species. On an even finer scale, consistent differences have been found between the cuticular hydrocarbons of races of various pest species such as fire ants and thus they have proved useful in identifying the origins of particular outbreaks.

The reason that cuticular hydrocarbons have such taxonomic potential is not pure chance but stems from the fact that in addition to contributing generally to waterproofing, they often play crucial roles in species, gender, and in the case of social insects, colony recognition. Thus in many respects they are subject to similar selective pressures as sex pheromones. They are not, however, immune from homoplasy, and in particular numerous instances have been found of convergent evolution (chemical mimicry) in cuticular hydrocarbon composition between inquilines, insects that live within nests of another species, and their hosts (Howard *et al.*, 1990).

8.2.3   *Secondary plant metabolites*

The field of secondary plant substances has been studied more intensely than almost any other aspect of chemotaxonomy in terms of their potential taxonomic significance. For many years the functions of most secondary plant substances were unknown but whilst much remains to be discovered it is becoming clear that collectively they perform a wide range of functions including photobiological roles, herbivore deterrence, anti-fungal, -bacterial and -viral activity, and pollinator attraction. Further, it should be remembered that flower coloration is a direct expression of the presence of particular chemicals. However, probably one of the main roles that these compounds assume is that of excretion, plants having the potential to package-up unwanted compounds in vacuoles in their leaves. In the case of deciduous plants, the shedding of old leaves loaded with especially deposited secondary plant compounds is akin to the excretory processes of animals. Hence the brilliance of autumnal colours.

Among the types of chemicals that have been most intensively investigated must be listed the flavonoids (Crawford, 1978), triterpenes, diterpenes, sesquiterpenes and their derivatives, and the alkaloids (Aynilian *et al.*, 1973). In each of these groups large numbers of distinct compounds have been identified in different plants. In the sesquiterpene lactones for instance, more than 1300 distinct structures have been identified in just one plant family, the Asteraceae (Seaman and Funk, 1983). Before discussing a few examples it is worth noting that unlike most morphological characters which once expressed may remain with the organism for the rest of its life, the expression of genes for secondary plant metabolites may in some instances be turned on and off according to environmental factors. Thus there is a possibility that failure to sample plants from a diversity of habitats and at different times of the year could result in missing certain compounds. For studies of these compounds to be valid therefore, it is desirable that at least some investigation of possible levels of interspecific variability should be conducted.

Probably the most important general group of compounds are the many and diverse phenolic compounds synthesized by plants. Of these, much work and many important findings have been derived from the polycyclic flavonoids including the anthocyanins which are responsible, among other things, for flower colours. Of the steroids, one group are known as saponins because of their soap-like properties (though some 'saponins' have non-steroid structures), while another important group are the cardiac glycosides (= cardenolides) which have powerful physiological effects on many herbivores. Saponins are of rather limited distribution and as such are likely to be phylogenetically informative. In addition, the presence of some saponins can be detected readily by their ability to lyse red blood cells (hemolysis). As demonstrated by Jurzysta *et al.* (1988), hemolytic activity provides a simple diagnostic test to distinguish between the vegetative parts of members of the alfalfa genus, *Medicago*, from those of related taxa.

8.2.4 *Neurotransmitters*

Most multicellular animals have specialized nervous systems in which signals between neurones are transmitted principally by chemical means. A great variety of neurotransmitter and neuromodulator molecules are known to occur in central nervous systems including a diversity of oligopeptides. Within the peripheral motor nervous system a rather more restricted range of transmitter molecules are known with acetyl choline being the principal excitant of skeletal muscles in many groups including the vertebrates. Interestingly, whereas acetyl choline is also widespread among the invertebrate phyla, the Arthropoda (viz. Chelicerata, Crustacea and Uniramia) are distinguished in having L-glutamic acid as their principle excitatory peripheral neurotransmitter. So far little work has been conducted on the phylogenetic significance of neurotransmitter distribution probably because the neurophysiologists responsible for gathering the necessary data generally have little interest in taxonomy, although a start has been made by Walker and Holden-Dye (1989). In the above example, if it turns out that glutamate as a peripheral excitatory neurotransmitter is apomorphous with respect to acetyl choline it could add substantially to arguments in favour of a monophyletic Arthropoda.

8.2.5 *Pigments*

Chemicals contributing to pigmentation belong to an enormously diverse number of classes offering a wealth of phylogenetic information. Perhaps one of the most obvious groups of organisms for pigment research are the brightly coloured butterflies and indeed here the work of Ford has shown that pigment chemistry has considerable phylogenetic significance. One example will suffice, namely the group known generally as 'the whites' which comprises the family Pieridae. This group mostly makes use of a class of pigments called pterins to form their white and yellow colours. However, the same effect is achieved by the use of white flavones in the unrelated but possibly mimetic marbled white butterfly, *Melanargia galathea*, a European species which belongs to the brown butterfly family Satyridae (Ford, 1957).

8.2.6 *Animal toxins*

Toxins as a source of taxonomic information are best treated as chemicals that can be extracted and analysed; however, their effects alone have been taken as characters in some cases. A diverse array of animals produce toxins for defence, offence and prey capture. In spiders, for example, all but one of the 50 or so families, the Uloboridae, relies on the use of venom to subdue their predominantly arthropodan prey (Quicke and Usherwood, 1990), and collectively these compounds, although clearly being directed at much the same function, display a great diversity of chemical structures from complex proteins to small novel sophisticated molecules.

Surprisingly, given their comparatively complicated nature involving aromatic, amino acid, aliphatic acid and polyamine residues, toxins of the orb web-weaving araneid spiders are quite closely mimicked by a toxin of similar function yet obviously independent evolution found in the bee wolf, *Philanthus triangulum*, a predatory wasp that as its name implies specializes in hunting honey bees (Figure 8.1; Quicke, 1988).

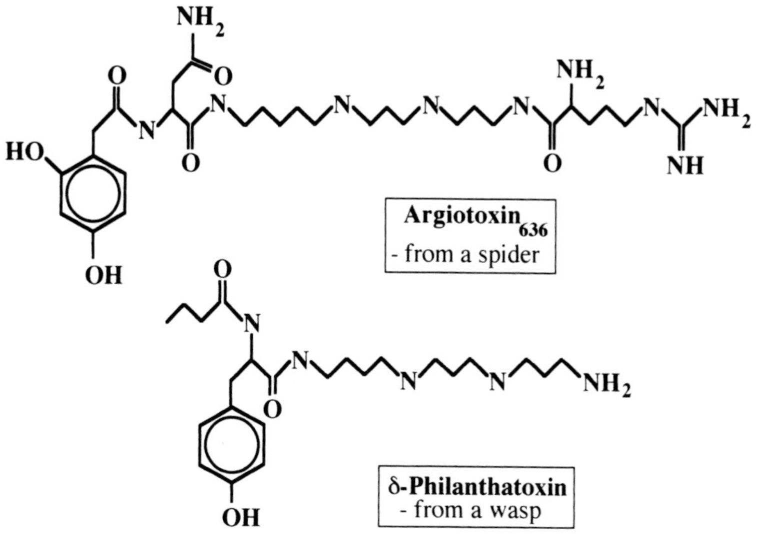

Figure 8.1 Two different toxins which block the excitation-mediating glutamate receptor at the insect nerve-muscle junction. δ-Philanthatoxin is produced by the bee wolf, a wasp that attacks honey bees and argiotoxin is produced by orb web-weaving spiders. The two toxins are undoubtedly convergent in structure with their combination of a phenolic residue (left) and long flexible positively-charged polyamine chains.

Both spider and wasp toxins target the same receptor molecule at their insect prey's neuromuscular junctions and evolution has solved the problem in very similar ways. Thus even sophisticated chemicals can show homoplasy at least in their general structural features.

8.2.7 Pyrolysis products

One of the more unusual sources of chemicals for use in taxonomic studies, particularly identification, is through **pyrolysis**, the degradation of biological macromolecules at high temperature in an inert atmosphere. This procedure yields an extremely complex mixture of smaller volatile components which can be separated by gas chromatography or more amenably, by mass spectroscopy. Its use in taxonomy is (thankfully) limited to microbiology where for some identification purposes it seems to offer a fairly robust means of distinguishing even between strains of a single species (Gutteridge *et al.*, 1985).

Essentially, the components of the mass spectrum of pyrolysis products (produced under standard conditions) are similar between taxa but their relative abundance varies significantly between taxa. Data obtained from unknown samples may then be compared using, for example, computerized discriminant techniques (see section 4.4), and an identification made.

8.3   Fermentation properties and drug resistance in microorganisms

For many years the identification of bacteria has relied on a small number of relatively simple observations and biochemical tests, paramount among which has been their ability to metabolize various compounds as energy sources. Most recently, drug resistance tests have also been immensely important in defining strains, many of which may have great medical or veterinary significance. Such biochemical tests are in fact indirect ways of determining whether an organism possesses one or more particular functional enzymes.

8.4   The use of chemical data

Assuming that a set of chemicals in a group of organisms have been identified, it is pertinent to ask whether these can be employed for the purpose of phylogenetic reconstruction. Several factors need to be taken into consideration, probably the most important of which is whether the expression of a particular chemical is potentially influenced by environmental or other factors such as an organism's age, sex, health, etc. Just as with different allelic variants of enzymes and other proteins dealt with in chapter 10, it is often the case that particular compounds may be produced only at certain times or under certain conditions. This results from the differential activity of genes encoding enzymes involved in particular pathways. It is therefore important before embarking on a chemosystematic study to try to maximize the chance that the organisms used will be comparable.

Several authors have tried to make biochemically logical decisions about the likely evolutionary pathways leading to a family of different chemical compounds. Any such hypotheses must necessarily be conjecture, and weak conjecture at that, since there may be many potential routes for the synthesis of one compound from another. To try and overcome this, some workers have therefore chosen to apply the principle of parsimony to such decisions in that they deem the most likely biosynthetic pathway between a set of compounds is the one that requires the fewest biosynthetic manipulations (Levy, 1977). This is analogous to the concept of the minimum number of bases in a nucleic acid that have to change for one amino acid to be replaced by another (see section 10.5.1).

It is probably true to say that in general, chemotaxonomy in spite of often high hopes, has tended to act very much as a supporter of classical morphotaxonomy when it comes to identifying evolutionary relationships. Only in comparatively few instances has it led to the realignment of higher taxa. Given that many chemicals investigated have highly adaptive roles to play in physiological processes, this is perhaps not overly surprising. Chemicals can often be synthesized by a variety of pathways and, like macromolecular sequences, it is impossible to distinguish homologous from analogous expression of a given compound without additional information. Nevertheless, chemotaxonomy has been remarkably successful in some areas. The obvious examples involve the identification of microorganisms in which the paucity of external morphological characters forces the taxonomist to search elsewhere for character systems, and to a lesser extent in lichens where chemical features have long played an important role in identification. Chemotaxonomy has helped in the recognition of species, sibling species and hybrids, especially among plants and fungi, but its use has overall been largely restricted to lower level problems (Harborne, 1968).

Given that many workers have undoubtedly felt that chemicals ought to provide strong taxonomic evidence it is pertinent to ask whether chemotaxonomy has let us down. The answer is probably yes, but the reasons may reflect our misinterpretation of what chemical data are. In fact, the presence of a particular chemical in an organism minimally only reflects the presence of one particular active gene for the enzyme required to synthesize it, though in many cases a more elaborate chemosynthetic pathway involving many enzymes might need to be in place (Stace, 1980). It may therefore be unreasonable to expect any single chemical feature to carry more weight than many morphological characters do. Much will depend on the particular chemical and circumstances in question. Further, chemicals can be subject to parallel evolution, convergence (see Figure 8.1) and reversals just like any other feature. Some self-deception on behalf of the chemotaxonomist may also have taken place because the accurate identification of a particular chemical may involve many laborious and expensive protocols compared with the simple viewing of a mor-phocharacter, and therefore for the costs sunk in making a chemical identification it is likely that we expect to get more than we are entitled to, in terms of phylogenetic inference.

CHAPTER NINE

IMMUNOTAXONOMY

One of the basic properties of the immune system is its specificity. A rabbit may contain up to 100 000 different kinds of protein molecules of its own. Yet, it will produce only antibodies against the foreign molecules with which it is immunised.

(Westbroek *et al.*, 1979)

9.1 History

The origins of the science of immunology date back to Pasteur who in the latter part of the nineteenth century first recognized that mammals could acquire immunity to pathogens by prior exposure to incapacitated or sublethal doses of the infectious microorganisms. The significance of this to taxonomy, as well as to medicine, lies in the fact that the acquired immunity is specific to the type of pathogenic organism used to provide the initial inoculum, and therefore the specific reaction of the inoculated animal is dependent on the chemistry of the inoculated components. Thus, the method can be used to detect the presence of the same or functionally similar antigens in different species.

Following Pasteur's original discovery, it was soon realized that the inoculum does not need to contain whole organisms, but that an immunological response can be achieved using purified macro-molecules, or mixtures of the same, although it took some while before details of the chemistry involved were worked out. Molecules or mixtures of molecules that can initiate an immunological response are called **antigens**. Any of a wide range of macromolecules can act as antigens including proteins, glycoproteins, carbohydrates, lipids and nucleic acids.

When samples of a foreign antigen are injected into a suitable immunocompetent animal, usually a horse, rabbit or rat, the immune system of the recipient animal will respond by producing specialized proteins (**antibodies**) which are capable of recognizing and binding specifically to the antigens. Several similar though distinct types of antibody are produced by the mammalian immune system, although collectively they comprise the **immunoglobulins** (IgGs) and can be fairly easily separated from other blood serum components. The principal antibody type from the immunotaxonomists's point of view are the IgGs which are produced in large quantities.

The amount of antibody produced by a single inoculation into a immunologically naive donor is generally rather small, so for practical purposes, antibody production is achieved by using a series of inoculations, the amount of antibody produced typically reaching a maximum after about four treatments and then remaining at a high level without further challenge, often for several years (depending on the antigens employed).

The reaction between antibody and antigen depends on the three-dimensional structure of the molecules involved. In the case of a proteinaceous antigen, the efficiency of its interaction with an antibody will therefore depend not only on the amino acid residues in direct contact with the antibody (which may themselves be widely distributed through the protein sequence) but may even be influenced by amino acid substitutions at more remote sites which nevertheless affect the protein's tertiary and quaternary structure. Thus the strength of a given immunological interaction will be determined by the 'sum' of many different amino acid positions (**antigenic determinants**). When an antibody is reacted with the antigens against which it was raised it is called a **homologous** reaction, whereas if the reaction involves antigens from a different taxon then it is referred to as **heterologous**.

From a practical taxonomic point of view, what is important is that the more amino acid replacements that occur in proteins as two taxa diverge, the fewer antigenic determinants they will have in common. Therefore immunological distances represent measures of the proportions of antigenic determinants recognized by antibodies raised against antigens of one taxon that are also recognized by the same antibodies in an equivalent sample of potential antigens from another taxon.

9.2   Precipitin reaction

A very important property of antibodies with regard to their use in taxonomy is their ability to produce an insoluble complex when mixed with antigen. This precipitate, called **precipitin**, forms when the concentrations of antibody and antigen reach optimal proportions. Typically these are optimally produced when the antibody to antigen ratio is about 1.5 to 1. In an excess of either antigen or of antibody the complexes formed remain soluble.

This reaction forms the basis of most of the techniques employed in immunotaxonomy with the exceptions of the microcomplement fixation (section 9.5) and radioimmunoassay (section 9.7). The precipitin reaction was first applied specifically to a taxonomic problem by Nuttall (1904) who with his co-workers used whole blood serum as their antigenic mixture and would thus have produced antibodies to a wide range of different immunogenic serum proteins and glycoproteins. Their results would therefore be expected to reflect both differences in the sequences of these proteins (through their ability to produce immunological cross-reactions) and differences in their relative abundances in the test sera. As such, their results would not have been 'clean' and nowadays similar studies are carried out in a far more rigorous fashion with purified and quantified protein extracts.

From a taxonomic view point, the results of precipitin tests can be interpreted either quantitatively or qualitatively depending on the protocol. In quantitative set-ups, the amount of precipitin formed under standard conditions is used directly as a measure of distance between the test organisms (**immunological distance**). These distance measures correlate in broad terms to the amount of amino acid sequence similarity between the different antigenic proteins and therefore it is possible to convert immunological distances to approximate numbers of amino acid substitutions.

9.3 Immunodiffusion

This standard but useful set of techniques is based on the precipitin reaction, which in this case takes place in a gel. The general method places an antiserum raised in an animal against a particular antigen (or antigen mixture) in a well cut into a thin layer of a suitable gel medium such as agar. Potential antigens, which in a taxonomic situation will probably be derived from different populations or taxa, are placed in one or more other sample wells, separated from the antiserum well by a few millimetres of gel (Figure 9.1). Both antibodies and potential antigens diffuse into the gel matrix and at a distance between the wells determined by their molecular size will come together in an appropriate ratio for precipitin formation. If the antibody recognizes antigenic determinants on the potential antigen molecules diffusing from a given well, an arc of precipitin will be formed and deposited in the gel, and this can then easily be visualized after washing unreacted proteins from the gel by using a general protein-staining procedure. If the antigen sample contains more than one antigenic component, which due to molecular size diffuse at different rates through the gel, several precipitin arcs will be formed at different positions between the antiserum and antigen wells. Conversely, if the antibodies in the antiserum do not recognize any of the antigenic determinants on the antigen samples, no precipitin will be formed (as with samples D–F in Figure 9.1).

In a commonly employed arrangement with different antigens placed

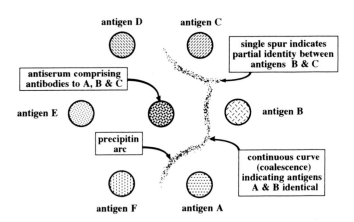

Figure 9.1 Stylized diagram of a double immunodiffusion gel (one of the arrangements proposed by Ouchterlony) in which antiserum (containing antibodies against different determinants of antigens A, B and C) diffuses from the central well and six different antigens diffuse from the outer ring of wells. Insoluble precipitin forms where concentrations of antibody and antigen reach optimal levels (soluble complexes are formed with excess of either antibody or antigen). The antigenic determinants of A and B are seen to be identical as the precipitin arcs formed by each coalesce perfectly. Antigen C contains an extra determinant recognized by some of the antibodies in the central well, not present in A or B, as is revealed by the presence of a spur of precipitin formation; in addition C also has some determinants in common with B as indicated by the partial coalescence of the precipitin arc between them.

in different but adjacent wells (Ouchterlony, 1953, 1958) it is possible to obtain even more information than whether or not each antigen reacts with the antiserum. Since antisera raised in animals are almost invariably polyclonal and the different antibodies that result are likely to recognize different antigenic determinants on the antigen molecule, it is possible for the antibody sample to recognize different antigens in the different sample wells. For example, in Figure 9.1 the continuous curve without any spurs formed by the precipitin arcs of antigens A and B shows that in terms of their antigenic determinants (as recognized by the antiserum in the central well) they are identical. However, the same is not true for antigens C and A (or B). Here a spur indicates that some precipitin is being formed closer to the antigen well, which must mean that some of the antibodies in the central well are not reaching optimal precipitin formation concentrations until they have diffused beyond the main precipitin band. Hence it must be that some of the antibodies in the central well are recognizing determinants on antigen C other than those present on either antigens A or B. Thus antigen C is partly but not completely homologous with antigen B. If two antigens, both of which are recognized by some antibodies in the central well but have no antigenic determinants in common, are placed in adjacent peripheral wells then the precipitin arcs will cross and there will be no sign of coalescence between them.

9.4 Immunoelectrophoresis

This useful system is a two stage technique in which pairs of soluble protein samples (antigens) are first separated along the length of a gel under standard electrophoresis conditions (Figure 9.2 (*left*); see chapter 10). After electrophoresis, the elongate central well is filled with antiserum raised against one of the antigen samples and the antigens and antibodies then diffuse through the gel until they meet. Since the antibody sample in the central well will normally be polyclonal, having been raised against a mixture of potential antigens such as in whole serum, the antibodies will usually form precipitin arcs where they meet a number of the separated antigenic proteins (Figure 9.2 (*right*)). Because the individual proteins in the original antigen sample have been spread out along the gel, the precipitin arcs formed

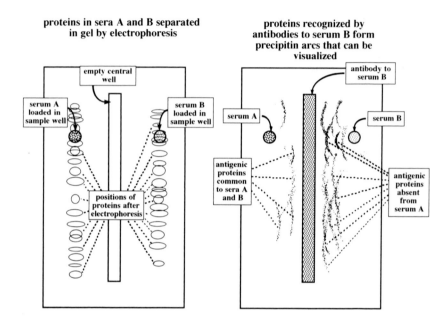

Figure 9.2 Diagram of an immunoelectrodiffusion gel plate for comparing antigens of two taxa. Following electrophoresis of the two antigens from the round wells, proteins (and glycoproteins) have migrated along the length of the gel (in both directions) (*left*). The central well is then filled with polyclonal antibody to serum B (originally loaded into the round right-hand well) which then diffuses laterally into the gel, and forms precipitin arcs where it meets appropriate concentrations of corresponding antigens (*right*). The results depicted show that antigen mixture A (originally placed in the left-hand well) contains five components that are immunologically similar to antigens in serum B.

with each protein will usually be distinguishable. This distinction is further helped because antigens of different molecular size will diffuse at different rates and therefore the precipitin arcs will not only be separated along the length of the gel but also along its width. On the side of the gel along which the heterologous antigen sample was run there will usually be fewer precipitin arcs formed because the more distantly related the source taxa are, the fewer antigenic determinants their proteins will have in common. Thus the difference in the number of precipitin arcs on the two sides of the gel compared with the maximum number will be a measure of the immunological distance between the two taxa.

9.5 Microcomplement fixation (MC'F)

This technique has been widely employed in phylogenetic studies because it allows a quantitative estimate to be made of the amount of sequence divergence in a particular antigenic protein. Unlike the immunodiffusion techniques described above, microcomplement fixation requires a purified protein, but only a relatively small amount so it can be quite economical with material, especially if an abundantly synthesized protein is selected for study. The underlying rational for the technique relies on the probability that any given protein will have quite a few separate antigenic sites (ideally at least 20). Therefore if a **polyclonal antibody** is raised against this purified protein in an inoculated animal, individual antibodies will collectively recognize and bind to a good proportion of its antigenic sites (Figure 9.3). Then, depending on the time that elapses after a pair of taxa diverge, an homologous protein in a related taxon will probably have lost some or all of these determinants. The more the amino acid sequences of the two proteins have diverged, the fewer antigenic determinants are likely still to be recognized by the polyclonal antibody mixture raised against one of the proteins. Thus, the amount of antibody binding per antigen molecule ought to provide a measure of overall sequence similarity. Importantly, MC'F is carried out at very low concentrations of antigen and antibody such that binding only occurs between an antibody and its corresponding antigenic determinant if the latter is virtually identical to the one the antibody was raised against. Even a single amino acid difference between two proteins that affects one of the antigenic determinants is likely to prevent antibody binding under MC'F conditions (Maxson and Maxson, 1990).

Microcomplement fixation provides a measure of antigenic similarity by comparing the concentration of antibody bound to the homologous antigen (i.e. the protein which was used to stimulate antibody production) with that bound to a heterologous one. To assay the amount of binding between known concentrations of antibody and antigen, a precise amount of **complement** is included in the reaction mixture. Complement associates only with tightly bound antibody–antigen complexes and therefore the amount of complement bound is proportional to the number of antigenic determinant sites that are tightly bound to an antibody in the test. Further, given that the initial concentration of complement is precisely known, then the amount that remains free in solution is inversely proportional to the number of

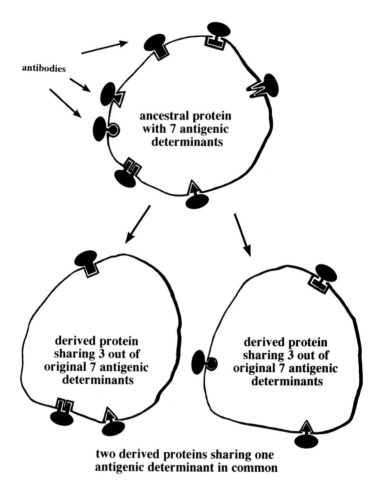

Figure 9.3 Principle underlying microcomplement fixation. A stylized antigen molecule with seven antigenic sites with their corresponding bound antibodies (*above*). Two evolutionary derivatives of the protein illustrated above which share some of the same antigenic determinants with the ancestral protein but fewer with each other (*below*).

antigenic determinants present. In practice, the free complement is then assayed by its ability to lyse sheep red blood cells that have been sensitized previously by coating them with anti-sheep red blood cell antibody.

For MC′F the antigenic proteins from two or more species must be purified and accurately quantified, and an antiserum raised against one or more of these. A set of standard complement fixation curves is then prepared for a range of homologous antigen and antibody concentrations so as to obtain a family of curves such as is shown in Figure 9.4. From these it is possible to determine the maximum amount of complement fixation that occurs with each antibody concentration, a value usually referred to as the **titre**. The same procedure is simultaneously carried out using the same antibody but with heterologous antigen. The difference in the concentrations of antibody required to produce identical maximum amounts of complement fixation with the two antigens is then used as a measure of the dissimilarity between the two antigens.

The potential of microcomplement fixation to provide intertaxon distances suitable for phylogenetic inference lies in how closely the relationship between antibody binding reflects the degree of sequence divergence between homologous and heterologous antigens. Assumptions that the relationship should be well behaved have been based largely on intuition rather than on hard fact. Nevertheless, use of an empirically derived logarithmic transformation of the absolute antibody concentration bound to the antigen as a basis of an immunological distance measure has been widely accepted, perhaps because these

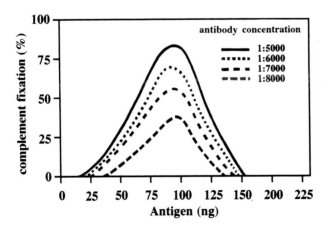

Figure 9.4 Set of four microcomplement fixation curves obtained with four different concentrations of antibody to the antigen. Immunological distance is determined by matching the amount of the same antibody that is required to produce equivalent complement fixation when reacted with the same concentration of an homologous antigen from a different OTU (operational taxonomic unit).

immunological distances have often seemed to agree rather well with other estimates of evolutionary distance.

The immunological distance, *ID*, can therefore simply be defined as:

$$ID = 100 \, Log_{10} \left(\frac{\text{heterologous antibody titre}}{\text{homologous antibody titre}} \right)$$

Under MC'F conditions a single amino acid replacement will usually be enough to prevent the antibody from binding to the antigenic determinant in its corresponding heterologous protein. Because of this, Maxson and Maxson (1986) were able to show that the value of *ID* is also directly related to the average number of amino acid replacements per antigenic determinant with the latter ratio being given by *ID*/43.

Serum albumins have been widely employed in MC'F studies of vertebrate relationships because they are easily purified, abundant, large and highly antigenic molecules with about 25 antigenic sites each. Similarly, antigenic proteins in insect haemolymph have proved suitable for MC'F (Beverley and Wilson, 1982) and so too have various easily purifiable enzymes. Detailed procedures for carrying out MC'F comparisons are described by Maxson and Maxson (1990).

9.6   Use of monoclonal antibodies

Until very recently, most taxonomic applications of immunology have used **polyclonal** antibody preparations, that is ones which potentially comprise a mixture of many different antibodies each recognizing different antigenic sites on the antibody molecule. Thus while conventional immunological results obtained using these mixtures will give a measure of overall similarity between two proteins, they cannot resolve variation at individual antigenic sites. The rapid development of techniques for preparing **monoclonal** antibodies, which recognize only a single antigenic determinant on the target molecule, radically changes the type of data that immunotaxonomy can provide. Instead of simply giving distance data, monoclonal antibodies, because of their specificity, make it possible to score the presence or absence of particular antigenic sites in proteins from different organisms and therefore this data can be treated more readily in parsimony analysis. As pointed out by Greenstone *et al.* (1991), production of monoclonal antibodies is both time-consuming and expensive and would therefore be an unlikely first choice for taxonomic work. However, as time goes by, more and more monoclonal antibodies will become available from chemical suppliers making it possible to buy ready-made 'probes' for this most sophisticated form of immunotaxonomy.

9.7 Radioimmunoassay

Radioimmunoassay is another fairly modern technique that is likely to have an increasing role in taxonomic studies because of its sensitivity which at best requires only picogram quantities of antigenic proteins. The basis of radioimmunoassay is that soluble antigens can be made to bind in proportion to their abundance to the sides of suitable sample wells in a polyvinyl plate. When bound, their exposed sides are still available for binding to antibody, and thus if a sample well with its bound antigens is exposed to antibodies that recognize these bound antigenic determinants, the antibody will also stick to the well wall forming a sandwich arrangement. The amount of antibody that binds can then be determined by reacting it with another antibody, this time a radiolabelled one that has been raised in a different species of animal against the constant region of the first antibody. Consequently a triple layer sandwich is formed. Typically the experimental antibodies will be raised in rabbits and these will be detected using ^{125}I-labelled antibodies raised in goats against rabbit τ-globulin (usually abbreviated as goat anti-rabbit τ-globulin).

The sensitivity of radioimmunoassay can make it the only choice for some projects where the amounts of antigen available from some taxa may be very limited. One such example is provided by the use of radioimmunoassay to detect antigenic components in fossil materials (Westbroek et al., 1979). However, against this has to be weighed the risks inherent in the use of radioactive iodine (^{125}I) and its not inconsiderable cost.

9.8 Analysis of immunological data

Virtually all immunological data is in the form of intertaxa distances and for the most part it is not possible to polarize individual traits in evolutionary terms. Distances can be calculated in a variety of ways using indices such as those presented in chapter 4. In the special case of MC'F immunological distance, as defined in section 9.5 above, is approximately linearly related to the number of amino acid replacements that have taken place. The quality of such distance data is best assessed using reciprocal assays, that is by determining the distance between two taxa using antigens and antisera from each. If some antisera are of poorer quality than others, a correction factor can be used in order to rescale the distance data.

As with other distance matrices, trees can be constructed from immunological distance data using any of a number of methods, for example, using phenetic clustering methods such as UPGMA (see section 4.3.3) or parsimony-based procedures such as the Fitch–Margoliash or distance Wagner methods (see section 3.2.3).

CHAPTER TEN

PROTEINS AND TAXONOMY

Almost as soon as macromolecules were sequenced, predictions were made that the sequences contained information about evolutionary history . . .

Penny and Hendy (1985)

10.1 Introduction

Since the amino acid sequences of proteins are encoded by the nucleotide sequences of a cell's DNA, variation in protein structure between organisms effectively provides a window into the cell's genome, and for many years provided one of the simplest means of obtaining information directly about gene sequences. However, proteins are complex molecules, and study of the amino acid sequences that make them up was until quite recently a difficult lengthy and costly procedure. Much of the usefulness of proteins in taxonomic investigations has been the result of investigations that compare the physical properties of particular proteins rather than working out their amino acid sequences. Paramount among the procedures employed for investigating proteins are a number of techniques generally referred to as **electrophoresis** several variants of which are described briefly below.

10.2 Techniques of protein electrophoresis

Electrophoresis relies on the fact that the protein or other molecules being separated can have a net electric charge when in solution. Specifically, any protein will have a net charge over a range of pH values due to the ionization state of the various ionizable side chains they possess. Soluble proteins can therefore be made to migrate within an electric gradient at a rate that depends upon their net electric charge and sometimes on their molecular size and shape. As slightly different forms of a given protein (having slightly different amino acid sequences) usually vary relatively more in charge than in molecular weight, the use of charge for separation is far more likely to reveal any variation present.

In practice a protein sample to be electrophoresed is placed at one end of a suitable medium, usually a thin slab of a gel or similar porous matrix, and an electric field is then applied across the length of the gel. The different soluble proteins in the sample will migrate at different velocities within the applied electric field according to their charge, molecular weight, etc. This relatively simple basic process has been responsible for major advances in many fields of biology, not least systematics. Commonly employed gel media include starch, poly-acrylamide and cellulose acetate.

The term electrophoresis does not refer to a single technique but rather covers a range of procedures. A brief outline of some of the more commonly encountered variants is given below. Modifications of the basic technique enable proteins and similar macromolecules to be separated specifically according to their folded size, their molecular weight, their net charge or their **isoelectric point**.

10.2.1   *SDS polyacrylamide electrophoresis*

SDS or **sodium dodecyl sulphate** is a simple ionic detergent which binds readily to protein chains at fairly regular intervals. Further, treatment of a globular protein with SDS causes it to unfold into a straight chain (except where the shape is held intact by disulphide bridges) and consequently destroys any biological activity such as enzymic function. However, because the negatively charged dodecyl sulphate residues attach at regular intervals along the now denatured protein, the overall charge of the protein–SDS complex will be determined effectively by the length of the original protein chain rather than by its amino acid composition. Electrophoretic separation of SDS-treated protein samples in an appropriate strength gel will therefore result in the proteins migrating at a rate closely related to their molecular weight.

10.2.2   *Gradient gel electrophoresis*

The movement of proteins through a gel under electrophoresis is ultimately limited by the size of the pores in the gel. If the pores are too small then the protein will be unable to migrate through them. Under normal electrophoresis, the gel strength is selected so that it does not prohibit the migration of any of the proteins of interest. However, by means of a simple mixing apparatus, it is possible to cast a polyacrylamide gel whose strength changes in a consistent manner from one end to the other. Such separation matrices are called **gradient** gels and they have the property that if the initial protein sample is placed at the low strength (large pore size) end, the macromolecules being electrophoresed will migrate through the gel at progressively reducing rates until they reach a gel concentration through which they can no longer pass. Therefore they end up being separated according to their physical size.

10.2.3 Isoelectric focusing

Ordinary electrophoresis can resolve differences between proteins that result from the molecules having amino acid substitutions which give them a different net charge. However, it is unable to distinguish between two nearly identical proteins that differ in an amino acid if the substituted amino acid has the same charge as the one it replaced. For example, a substitution in which a lysine residue (having an amino group in its side chain) is replaced by a glutamic acid residue (with its carboxylic acid side chain) will give the protein a different net charge and should be readily separable by electrophoresis; however, if instead the lysine is substituted by an arginine which also has an amino group in its side chain, both normal and substituted proteins will have the same net charge under most circumstances and so will be indistinguishable. As it is likely that substitutions involving functionally important residues will involve amino acids with similar side chain properties, it is to be expected that the enzyme polymorphisms detected by ordinary electrophoresis are a considerable underestimate of the total number, many presumably having gone undetected. In fact it has been estimated that standard electrophoresis will reveal only one-third of the total number of amino acid substitutions in a protein.

Isoelectric focusing (IEF) provides a simple, although more expensive method capable of detecting differences between proteins involving substitutions between chemically similar amino acids that do not necessarily affect the net charge of the protein. Many of the various amino acid side chains of a protein will be ionized at any given pH value, and therefore the whole protein is likely to have a net charge too. However, amino acids and proteins which have both carboxylic acid and amino groups are known to chemists as zwitterions. Changing the pH environment of a zwitterion and hence of a protein, will change the ionization states of its component carboxylic acid and amino groups such that at low pH only the amino groups will be ionized and at high pH, only the carboxylic acid groups. The conditions for ionization of each of the potentially charged groups of a protein will depend on the particular amino acids concerned and on their local molecular environments. It follows that there will be a range of pH values over which some, but not all, amino or carboxylic acid groups will be ionized and therefore there will be a pH at which a protein will have an equal number of positively and negatively charged groups. At this pH, it will have no net charge. This pH is called the **isoelectric point** of the protein.

The isoelectric point of a protein is sensitive to even small differences in amino acid composition such that it can readily distinguish between proteins which differ in substitutions involving similarly charged amino acids or even in substitutions involving uncharged residues, as the latter may still be capable of influencing the ionizability of adjacent charged residues.

The crucial step in the realization of protein separation according to their isoelectric points is the construction of stable pH gradients. These are now routinely achieved using moderate molecular weight zwitterion materials called ampholytes which have considerable buffering qualities. Ampholytes, when incorporated into a supporting gel matrix, form a stable pH gradient when subjected to a strong electric field with the cathode having a high pH and the anode a low one. Given this gradient, samples of proteins to be separated can now be placed anywhere on the gel and they will migrate either towards the anode if they are initially placed on the basic side of their isoelectric points, or towards the cathode if they are placed on the acid side. Eventually the proteins will arrive at the pH where they carry no net electric charge and so will move no further. They are then said to have been focused. In practice it is possible to separate proteins differing in isoelectric point by as little as 0.02 pH units.

However, isoelectric focusing should not necessarily be seen as a panacea replacing standard electrophoresis. Several recent studies have shown that while IEF can resolve differences missed by standard electrophoresis it may miss others and overall it may not offer a great deal of advantage over standard electrophoresis (McLellan and Inouye, 1986). For maximum discrimination IEF should be considered as an adjunct to standard electrophoresis rather than a replacement.

10.2.4 *Two-dimensional electrophoresis*

The mixtures of proteins obtained by homogenizing a tissue sample are often so complex that simply separating them by electrophoresis will not enable the vast majority of components to be distinguished. Instead many different proteins will be effectively 'hidden' in a smear of overlapping bands. One way of overcoming this is to separate the mixture in two dimensions using different procedures for the two separations (Figure 10.1); if the same procedure were used for both separations the proteins would merely become arranged along the diagonal with no increase in resolution. Several variants on this idea have been more or less widely employed in systematic and related studies:

- standard electrophoresis followed by IEF
- standard electrophoresis followed by SDS electrophoresis
- IEF followed by SDS electrophoresis
- standard electrophoresis followed by gradient electrophoresis

In each case, following the first separation, the strip of gel containing the electrophoresed proteins is then cut out and placed along the top edge of a second gel prior to the second electrophoretic step, which this time will be carried out at right angles to the first separation (Figure 10.1). Of the above techniques, only the first two allow an examination of enzyme activity since SDS treatment denatures proteins causing them to lose their activity. Two-dimensional electrophoresis has been widely used in the past to examine complex protein mixtures. In these cases, the proteins are identified virtually entirely on the basis of the positions of their spots on the gel and consequently there is a finite chance that proteins that appear similar in this way could in fact be quite different. Also, while it is not strictly speaking necessary, such data are usually analysed in the form of intertaxon distance data rather than discrete characters.

10.3 Systematic aspects of electrophoresis

While it is possible to obtain useful systematic information from electrophoretic analysis of crude biological protein samples, the large numbers of different soluble proteins in an average tissue homogenate means that simple electrophoretic separation usually yields a largely

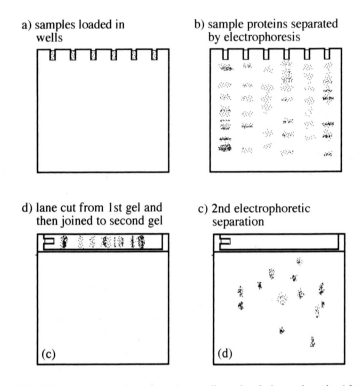

a) samples loaded in wells

b) sample proteins separated by electrophoresis

d) lane cut from 1st gel and then joined to second gel

c) 2nd electrophoretic separation

Figure 10.1 Diagram showing four stages in two-dimensional electrophoresis. After the first electrophoretic separation, one of the lanes along which sample components have been separated is cut from the gel and fused sideways-on with a new gel usually using more gel medium. The second separation, carried out at right angles to the first but under different conditions, often reveals that bands seen in the first separation actually comprise more than one component whose mobilities differ under the two different separation conditions.

uninterpretable array of protein bands. This problem is even more acute with SDS separations, first because SDS will frequently solubilize otherwise insoluble proteins adding to the number of bands, and second because SDS gels only give information on protein molecular weight rather than on native charge. Nevertheless, in a few cases comparisons between species and even populations of a single species have proved both possible and rewarding. For example, Coats *et al.* (1990) studying the parthenogenetic rose beetle, *Asynonychus godnami* showed qualitative differences in soluble protein composition between populations from different parts of the USA.

A major breakthrough in the use of electrophoresis in systematics came with the discovery that special histochemical staining procedures previously used to reveal sites of activity of specific enzymes in tissue sections could be adapted to reveal the locations of electrophoresed bands of the same enzymes in gels. In this way, instead of seeing all the proteins that were present in the initial sample, only a very small number of bands, corresponding to the positions of bands of the particular enzyme in the gel, will be apparent. Electrophoretic gels stained for specific enzyme activities are often referred to as **zymograms**.

10.3.1   *Isozymes and allozymes*

Enzyme electrophoresis is capable of revealing two distinct types of genetically controlled variation in enzyme phenotype. First, for any gene locus coding for a particular enzyme protein, there may be more than one different allele, each coding for a slightly different amino acid sequence, which in turn causes a difference in mobility during electrophoresis. These enzyme variants at a single locus are termed **alloenzymes** or **allozymes** and electrophoretic studies of their variation are termed **allozyme electrophoresis**. Second, the number of gene loci present in an organism that code for a protein with a particular enzymic activity may vary. For many simple organisms such as bacteria there may be only one locus for a given enzyme but in higher eukaryotes there is often more than one locus within the haploid genome. The enzyme products formed from genes at different loci are termed **isozymes** or **isoenzymes**.

Nowadays, a large number of different enzymes can be stained in gels including many that are involved in fundamental cellular processes. Important amongst these are the enzymes involved in glycolysis such as hexokinase, phosphofructokinase, phosphoglucomutase, aldolase, glycerol-3-phosphate dehydrogenase and lactate dehydrogenase, those involved in the Krebs citric acid cycle including aconitase, isocitrate dehydrogenase, malate dehydrogenase and succinate dehydrogenase, and the enzymes involved in protecting cells from free radicals, catalase and superoxide dismutase. These and a number of other enzymes are particularly useful because their virtually ubiquitous distribution among living organisms and their general presence at reasonable levels of activity means that similar detection techniques can be employed for almost any given tissue sample. Staining protocols may even be identical across kingdoms and usually only the choice of electrophoresis buffer systems may need to be optimized.

Those interested in carrying out allozyme electrophoresis can find many recipes for gel buffers and specific stains in the works of Shaw and Prasad (1970), Harris and Hopkinson (1976 *et seq.*), Richardson *et al.* (1986) and Murphy *et al.* (1990). Enzyme bands can be stained after electrophoresis either in normal gels or in isoelectric focusing gels. The latter being especially useful when dealing with minute samples because of the way proteins become concentrated at their isoelectric points (Kazmer, 1991) rather than becoming more spread out due to diffusion.

10.3.2 *Interpreting allozyme banding*

Once allozymes have been separated and stained they have to be interpreted. Firstly, allozyme banding patterns will depend on whether the active enzyme is monomeric (composed of only a single protein chain) or oligomeric, usually comprising two or four subunits. If a given enzyme is monomeric then the zymograms will normally show single bands of staining from homozygous individuals and a pair of bands for heterozygotes (Figure 10.2a), one band corresponding to the protein product of each of the two alleles possessed by the heterozygote. However, if the enzyme is a dimer (i.e. it comprises an association of two separate protein molecules coded by the same gene) then homozygotes will still yield only one enzyme band but heterozygotes will be represented by three bands (Figure 10.2b). The three-banded nature of dimeric enzymes from heterozygotes is explicable because with two different gene products it is possible to assemble three different protein dimers, one comprising two fast-migrating **electromorphs**, one comprising two slow electromorphs, and one comprising one of each. For trimeric proteins (which are quite rare) heterozygotes give four bands, for tetrameric enzymes (commoner) heterozygotes give five bands, etc.

At this point it should be emphasized that many enzyme systems often do not form clean cut, single bands even in homozygotes. Often two or more bands may be present due to modifications to some of the protein molecules that take place after they have been synthesized.

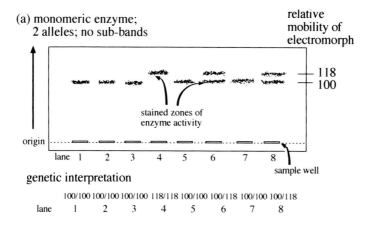

(a) monomeric enzyme;
2 alleles; no sub-bands

relative mobility of electromorph

118
100

stained zones of enzyme activity

origin

lane 1   2   3   4   5   6   7   8

sample well

genetic interpretation

100/100 100/100 100/100 118/118 100/100 100/118 100/100 100/118
lane   1   2   3   4   5   6   7   8

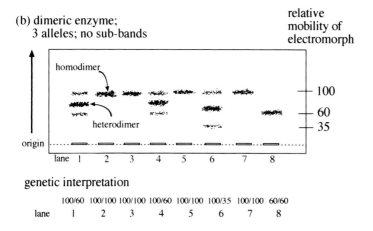

(b) dimeric enzyme;
3 alleles; no sub-bands

relative mobility of electromorph

homodimer

100

60

35

heterodimer

origin

lane 1   2   3   4   5   6   7   8

genetic interpretation

100/60 100/100 100/100 100/60 100/100 100/35 100/100 60/60
lane   1   2   3   4   5   6   7   8

Figure 10.2 Stylized diagram of two enzyme electrophoresis gels stained for a monomeric enzyme (a) and a dimeric one (b). In the case of the monomeric enzyme, two electromorphs are apparent, the faster migrating under these electrophoresis conditions, 1.18 times faster than the slower allozyme; samples from heterozygous individuals display two more or less equally staining bands. With the dimeric enzyme, three electromorphs are shown; samples from heterozygous individuals display three bands with the central one approximately twice as intensely staining as either of the other two. Dimeric enzymes of heterozygotes occur in three forms, one comprising two copies of the slower electromorph, one comprising two copies of the faster one, and one comprising a slow and a fast electromorph subunit. In a randomly associating mixture of allozyme subunits there will be twice as many heterodimers as either homodimer.

These variants can be an intrinsic aspect of the biochemistry of the organism or they may result from changes (e.g. oxidation) that can occur during storage of the protein sample. In the latter case, it may possible to reconvert some of the extra electromorphs back to the original form by including a reducing agent (e.g. mercaptoethanol or dithiothreitol) in samples and buffers during electrophoresis.

A different form of genetically controlled variation involving enzyme-coding genes is the null allele. That is some alleles for a given enzyme may give rise to an enzyme with no (or negligible) activity. In these cases a homozygote for the null allele will show no staining at the normal position on a zymogram while heterozygotes may show full or reduced activity depending on whether or not dosage compensation occurs. However, such null alleles are not easy to verify as such since the amount of enzyme activity detected in a gel (or present in an organism) can vary widely due to other factors.

10.3.3   *Analysis of allozyme data*

Electrophoretic protein separation, and allozyme analysis in particular, has had an outstanding impact on our understanding of many areas of biology including population genetics and taxonomy. For example, within the field of population genetics the ability to identify gene frequencies makes it possible to reveal whether two populations constitute a single gene pool or whether they are isolated and thus may need to be considered as separate species (**cryptic species**). At a different level, allozyme data can be used for constructing phylogenetic hypotheses at the species level and occasionally (e.g. among birds) at the generic level.

Since the enzyme banding phenotypes generally correspond to genotypes,* investigation of deduced gene (allele) frequencies for suitable polymorphic enzymes can reveal whether or not a series of samples were drawn from a single interbreeding population. In an interbreeding population with **panmixis** and in the absence of

---

* It must be emphasized, however, that there are plenty of other sources of variation in allozyme mobility that are not genetically controlled including phenotypically induced changes or chemical changes due to sample preparation or storage. In Flowerdew and Crisp's (1976) study of barnacles, for example, variation was found to occur with age, season and ecology. Therefore, before apparent variation is treated as genetically controlled, a number of checks need to be made to eliminate other possibilities.

selection, the frequencies of the different alleles of a polymorphic gene will be predicted by the **Hardy–Weinberg equilibrium**. However, if the sample from which allele (or corresponding allozyme) frequencies are drawn comprises two or more genetically isolated populations, allele frequencies will usually depart from Hardy–Weinberg equilibrium. Gene frequency information together with analysis of linkage can therefore be used to assess the numbers of separate genetically distinct populations (potentially separate taxa) represented in a sample of organisms. In practice, such allozyme analyses make use of information about as many different enzymes as possible, and if departures from Hardy–Weinberg equilibrium are detected care should be taken to ensure that these are not the result of some systematic error (Murphy *et al.*, 1990).

Soon after their discovery, isozymes were also being used in attempts to reconstruct phylogenies either within or between species. However, as personal computers were not available in those days and methodologies for parsimony analysis had hardly begun to be developed, early attempts mostly relied on converting isoenzyme data into distance or similarity measures, a survey of some of which is provided by Avise and Aquandro (1982). Probably the most widely used measures are the genetic distance of Nei (1972; and a modification for small samples in 1978) and the genetic similarity index of Rogers (1972).

Since the mobility of a particular protein depends largely on its net charge, isozymes that differ by two amino acids of similar charge will display a greater difference in mobility than if there was only a single charged amino acid difference between them. Given this, Krimbas and Sourdis (1987) have argued that isozyme mobility actually provides more data than simply whether two isozymes are likely to be the same, specifically it provides evidence of exactly how similar they are in terms of net charge. As a result they provide a modified form of genetic distance estimator which takes into account the degree of mobility variation between the alleles. However, independent phylogenies generally suggest that difference in allozyme migration speed is a poor estimator of underlying genetic difference.

More recently, a different approach to phylogenetic analysis based on allozyme data has taken into account each isozyme as a separate character with its particular set of allozymes as the character states. Such characters are treated as polymorphic and can be analysed by a variety of techniques (Swofford and Berlocher, 1987; sections 2.2.5, 2.2.6).

10.3.4 *Allozymes at subspecies, species and genus level*

Not long after isoenzyme analysis was developed, some workers took an interest in what this independent method might tell us about how taxonomists had been classifying their species. Among the questions that isozymes might be able to answer were: Just how different are species at the protein level? Are taxonomists generally consistent in the levels of divergence they interpret as representing species and genus level boundaries? Potentially can genetic distances derived from allozyme frequency data be used to determine, in a more rigorous way, the classification of taxa into species, genera, etc.?

Avise (1974) drew attention to the considerable potential of electrophoretic isoenzyme data in systematics research when he showed that measures of genetic similarity within species were generally extremely high over a large range of organisms while similarities between congeneric species gave much lower values with virtually no overlap between the two sets of results (Figure 10.3) although some differences are observed in the levels of genetic differentiation observed between congeneric species in different major groups of organisms (Figure 10.4). In fact, species within a genus are often completely distinct in the alleles they possess at between 20% and 80% of their allozyme loci. Such levels of differentiation make isozyme analysis a potentially very useful tool for assessing gene flow between populations or between putative species and for reconstructing evolutionary histories. Jelnes (1986) considered at some length what allozyme data are needed in order to be able to sink species into synonymy and concluded that the only valid way involved collecting data on more than one enzyme for each individual sampled.

10.4 Chemical protein analysis procedures

The development of techniques that enabled fairly quick and easy protein sequencing for a time revolutionized the acquisition of data for taxonomic studies. Protein sequencing is still useful but is now rapidly

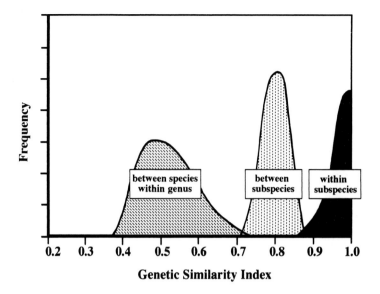

Figure 10.3 Distributions of genetic similarities between individuals within a subspecies (black), between subspecies within a species (vertical dashes) and between species within a genus. Note the virtual complete lack of overlap between the three classes of relationships which were themselves derived independently of allozyme data. (Data based on allozyme frequencies in sunfish are redrawn and stylized from Avise, 1975.)

giving way to nucleic acid sequencing procedures (see chapter 11) and therefore only a few relevant methods will be briefly described here.

10.4.1 Selective cutting of protein chains

An important group of proteases are the endopeptidases which can cut proteins at any position along their length where a particular amino acid or group of amino acids occur. Endopeptidases have therefore gained widespread usage in protein structure analysis because they can cleave large proteins into a number of more manageable smaller peptides that can be either electrophoresed directly or sequences completely. Among several endopeptidases in use are *Staphylococcus aureus* V8 protease which cuts proteins to the carboxyl side of glutamyl or glutamyl and aspartyl residues depending upon conditions (Table 10.1).

In addition to enzymes, some chemicals can also cleave peptide

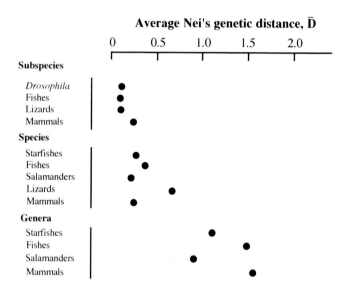

Figure 10.4 Average values of Nei's genetic distance based on allozyme frequency data shown separately for different levels of taxonomic separation in particular groups of organisms. Total ranges for each value are often rather large. (Data redrawn from Bruce and Ayala, 1979.)

chains at particular residues. Cyanogen bromide, for example, is an important reagent in protein chemistry because it selectively cuts proteins at the carboxyl side of methionine residues through its formation of a sulphonium salt with this amino acid. The usefulness is increased because methionines are generally rather rare in proteins so the action of cyanogen bromide usually produces a few large peptides rather than many smaller ones.

10.4.2   Chromatography of digests

The development of techniques for electrophoresing whole proteins lagged behind those suitable for smaller molecules because they required rather more sophisticated media than simple chromatography paper. One way around this problem, which yielded considerable data for taxonomy, was partially to digest purified proteins using proteases that cleave their substrates at particular amino acid sequences (Table

10.1). Trypsin, for example, cuts proteins selectively at arginine and lysine residues and so produces a number of smaller peptides from an original larger protein. These can then be separated by two-dimensional chromatography to give a characteristic pattern of ninhydrin-staining spots. Tryptic digests of similar proteins will usually have some peptides in common and some different, and these data can be employed in taxonomic studies. Nowadays, the analysis of digests has been superseded to a large extent by protein and gene sequencing although it is still used from time to time by less well-funded laboratories.

Table 10.1 Selectivity of some proteases used for protein cleavage

Enzyme	Cuts at
Trypsin	Carboxyl side of arginine and lysine
α-Chymotrypsin	Carboxyl side of leucine, methionine, phenylalanine, tryptophan and tyrosine (occasionally also at other sites)
Pepsin	Amino and carboxyl sides of leucine, methionine, phenylalanine, tryptophan and tyrosine
V8	Carboxyl side of glutamic (and aspartic) acid

10.4.3 *Amino acid sequencing*

Many proteins are produced by cells in reasonable amounts so that they can be extracted in large enough quantities for direct chemical analysis (i.e. amino acid sequencing). Further, as quite a large number of proteins, and particularly those involved in fundamental metabolic processes, may be found in a wide range of organisms, sequence analysis is potentially able to provide phylogenetic clues about many levels of relationship from species to family, order and even higher level. Notable amongst those proteins studied in this way are the cytochromes (particularly cytochrome C), albumins, haemoglobins and myoglobins.

For quite a long while, determining the sequence of amino acids in a protein offered the only practical way of inferring the sequence of bases in the DNA of the corresponding gene, albeit that there is a considerable degree of redundancy in the genetic code (see Table 10.2). Nevertheless, for many years protein sequencing was a long and

Table 10.2 Universal genetic code for the twenty amino acids occurring in proteins*

Amino acid	Abbreviation (single letter)	Property of side chain	mRNA triplet codes (codons)					
Alanine	A	hydrophobic	GCU	GCC	GCA	GCG		
Arginine	R	basic	CGU	CGG	CGA	CGC	AGA	AGG
Asparagine	N	hydrophilic neutral	AAU	AAC				
Aspartic acid	D	acidic	GAU	GAC				
Cysteine	C	hydrophilic neutral	UGU	UGC				
Glutamic acid	E	acidic	GAA	GAG				
Glutamine	Q	hydrophilic neutral	CAA	CAG				
Glycine	G	hydrophobic	GGU	GGC	GGA	GGG		
Histidine	H	basic	CAU	CAC				
Isoleucine	I	hydrophobic	AUU	AUC	AUA			
Leucine	L	hydrophobic	UUA	UUG	CUU	CUC	CUA	CUG
Lysine	K	basic	AAA	AAG				
Methionine	M	hydrophobic	AUG					
Phenylalanine	F	hydrophobic	UUU	UUC				
Proline	P	non-polar	CCU	CCC	CCA	CCG		
Serine	S	hydrophilic neutral	UCU	UCC	UCA	UCG	AGU	AGC
Threonine	T	hydrophilic neutral	ACU	ACC	ACA	ACG		
Tryptophan	W	hydrophilic neutral	UGG					
Tyrosine	Y	hydrophilic neutral	UAU	UAC				
Valine	V	hydrophobic	GUU	GUC	GUA	GUG		

* Based on the genetic code of cell nuclei of the majority of eukaryotes; organelle codes and some organisms use codes which differ to a greater or lesser extent from this, for example, yeasts and the protists *Euglena* and *Tetrahymena*.

arduous job requiring substantial quantities of purified protein and taking months or even years to sequence even a small number of residues. Fortunately, things have changed considerably in recent years and it is now possible to sequence proteins available in only milligram quantities in a matter of days or a few weeks. However, even these advances are rapidly being overwhelmed by advances in research into nucleic acids.

10.5   Analysis of amino acid sequence data

As with other sequence data, amino acid sequences have been analysed from a phylogenetic point of view either as overall distances or using each amino acid position as an individual character. The latter, however, is less strong than might at first appear and several relevant considerations will be dealt with below.

10.5.1   *Minimum nucleotide replacement*

As there are 20 different amino acids present in proteins that are directly coded for by messenger RNAs, it might seem reasonable when using protein sequence data in phylogenetic analysis to consider each amino acid position as a character that can assume any of 20 different states or, more precisely, 21 as amino acids might also become lost, and from this stance, to assume that any evolutionary transition from one amino acid to another is just as likely as any other. However, in terms of the number of distinct evolutionary steps that need to occur to replace one amino acid by another, transitions between different pairs of amino acids are not equal. This inequality results from differences in the number of positions in the triplet code that need to be changed in order for the new triplet to code for the new amino acid. For example, a change from methionine to tyrosine or vice versa necessitates changes at all three positions in the codons, whereas transitions between asparagine and lysine only require a change at the third position (Tables 10.2 and 10.3). Because of this, Fitch and Margoliash (1967) proposed that transitions between amino acids should be weighted in accordance with how many base changes in the codon are necessary to achieve the change between this pair of amino acids;

Table 10.3 Minimum number of nucleotide changes in a codon leading to a given amino acid substitution in a peptide. The full matrix is symmetric across the prime diagonal. Amino acids are abbreviated by the single letter code as shown in Table 10.2

	V	Y	W	T	S	P	F	M	K	L	I	H	G	Q	E	C	D	N	R	A
Alanine (A)	1	1	2	1	1	1	2	2	2	2	2	2	1	2	1	2	1	2	2	0
Arginine (R)	2	2	1	1	1	1	2	1	1	1	2	1	1	1	2	1	2	2	0	
Asparagine (N)	2	1	3	1	1	2	2	2	1	2	1	1	2	2	2	2	1	0		
Aspartic acid (D)	1	1	3	2	2	2	2	3	2	2	2	1	1	2	1	2	0			
Cysteine (C)	2	1	1	2	2	2	1	3	3	2	2	2	1	3	3	0				
Glutamic acid (E)	2	2	2	2	2	1	3	2	1	2	3	2	1	1	0					
Glutamine (Q)	2	2	2	2	2	1	3	2	1	1	3	1	2	0						
Glycine (G)	1	2	1	2	1	2	2	2	2	2	2	2	0							
Histidine (H)	2	1	3	2	2	1	2	3	2	1	2	0								
Isoleucine (I)	1	2	3	1	1	2	1	1	2	1	0									
Leucine (L)	1	2	2	2	1	1	1	1	2	0										
Lysine (K)	2	2	2	1	2	2	3	1	0											
Methionine (M)	1	3	2	2	2	2	2	0												
Phenylalanine (F)	1	1	2	1	1	2	0													
Proline (P)	2	2	2	2	1	0														
Serine (S)	2	1	1	1	0															
Threonine (T)	2	2	1	0																
Tryptophan (W)	2	2	0																	
Tyrosine (Y)	2	0																		
Valine (V)	0																			

specifically, the minimum number of changes that are necessary on the basis that any other assumption would be less parsimonious.

In practice, minimum nucleotide replacement data tables can be used in two forms of protein sequence analysis, distance-based analyses and parsimony analysis. In the former, the total sum of minimum nucleotide replacements is calculated by pairwise comparison between sequences to yield a distance matrix (Fitch and Margoliash, 1967). All the individual differences between two sequences are therefore reduced to a single distance measure. The alternative method treats each amino acid position as a separate character which can assume 21 different states (including deletions), and then subjecting this to standard parsimony analysis (see chapter 3). However, since parsimony analysis requires consideration of the likely sequences of the unknown ancestors, this has a slight but important effect on the way that the minimum nucleotide replacement matrix is used. Two factors in particular are worthy of note. Firstly, whereas differences in protein length can be allowed for quite simply in distance techniques by multiplying distances by a simple correction factor, parsimony analysis may either code missing characters as the loss of three nucleotides (thus weighting it by three) or as missing data (i.e. no extra length will be added). Of these two alternatives, the second is probably the best to assume as loss of a single nucleotide would cause a frame shift and all subsequent triplets would code for nonsense. Therefore, when codons are lost from functional genes it must usually result from a single event or a combination of events that remove a multiple of three bases.

Reconstruction of ancestral states with the parsimony method is further complicated due to the fact that unlike other amino acids, the codons for serine fall into two separate groups (group 1: UCU, UCG, UCA and UCC; group 2: AGU and AGC), and consequently distances for amino acid transitions involving serine will differ according to which codon may be involved at a given locus. Therefore, in parsimony analysis, it is necessary to treat serine as two separate characters and to select between these on the basis of the most parsimonious option.

The assumption that parsimony analysis of protein sequences is a valid way of estimating phylogeny has been elegantly tested by Penny and Hendy (1985) who compared the lengths of the shortest trees obtained with parsimony analysis of one protein sequence with the lengths of the same tree when measured using known sequences for other proteins. Their results are summarized in Figure 10.5 which shows that while there is generally a good correlation between tree lengths measured with different proteins, for no protein was the shortest (most parsimonious) tree also the shortest for any of the other proteins, nor were any of the shortest trees obtained from a single protein the same as the shortest tree obtained by analysis of the combined sequence data for all five proteins studied. Thus while it seems that parsimony analysis of protein sequence data may yield approximations of the true phylogeny, it may well be insufficient on its own to come up with the one true tree, and therefore other potential complicating factors may well need to be considered.

Of course, changing a single amino acid for another in a functional protein may lead to deleterious changes such as reduction or loss of enzymic activity or some equally undesirable effect. Evolutionarily

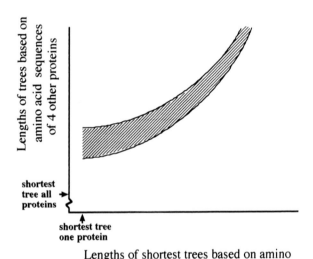

Lengths of shortest trees based on amino
acid sequence of one protein

Figure 10.5 Relationship between lengths of a set of parsimonious trees constructed using the amino acid sequence of one protein and the lengths of the same trees measured using amino acid sequence data from four other proteins (based on Penny and Hendy, 1985). The conclusion is reached that the shortest tree(s) found using any one protein is unlikely to be the shortest tree based on any other proteins or on all protein sequences combined.

successful transitions are therefore likely to be constrained. For example, substitutions of amino acids with acidic side chains by basic ones, or hydrophilic ones by hydrophobic ones are less likely to be successful than transitions involving replacements by amino acids with similar side-chain properties. However, assessing the degree of weighting that ought to be given to this is virtually impossible without far greater knowledge of the functions of particular residues. Further, a reanalysis of the minimum nucleotide replacement figures shows that transitions between amino acids possessing side chains with similar physicochemical properties, i.e. replacement of a hydrophobic residue by another hydrophobic one, require on average fewer nucleotide substitutions than do transitions between amino acids with different properties (Table 10.4). That a relationship exists at all is quite impressive given that the numbers of triplet codes per amino acid varies by a factor of six. The principal exceptions to this are transitions within the neutral hydrophilic groups, but this is not surprising since these amino acids are very diverse, including for example aromatic and non-aromatic compounds, and it may be presumed that they play a similar diversity of roles in protein function.

Following on from the above is the problem that for amino acid changes involving minimum nucleotide replacement values of two or more (Table 10.3) account ought to be taken of the natures of the intermediate amino acids. Thus as can be seen from Figure 10.6 both possible minimum length (2 step) pathways from the hydrophilic amino acid threonine to a hydrophilic tyrosine go via the similarly hydrophilic residues serine and asparagine. In contrast, a change from hydrophobic

Table 10.4 Mean numbers of nucleotide changes leading to a change in amino acid residue classified according to the physicochemical properties of the amino acids involved*

	Hydrophobic	Hydrophilic neutral	Acidic	Basic
Non-polar	1.71	1.5	2.0	1.33
Basic	1.86	1.57	1.67	1.33
Acidic	1.78	1.92	1.0	
Hydrophilic neutral	1.77	1.67		
Hydrophobic	1.38			

* The only non-polar amino acid is serine and so its distance to itself is not given.

alanine to hydrophobic methionine can go either via hydrophobic valine or via hydrophilic threonine. It seems reasonable to consider the route via threonine to be relatively disadvantageous, but by just how much?

A further, analogous complication arises from the fact that changes between codons for some amino acids can run via the stop codons UAA, UAG or UGA. For example, a change from lysine (AAG or AAA) to tyrosine (AAU or AAC) can go either via asparagine (AAU or AAC) or if the first base position changes first, through the stop codes UAA or UAG. Premature termination of a functional peptide may well result in a non-functional product and thus mutations leading to such an aborted translation might be strongly selected against. It might therefore be reasonable to consider weighting transitions such as between lysine and tyrosine as being only half as likely as some others.

Finally, the situation may be even more complex because in virtually all organisms there is a marked bias in the usage of synonymous codons for almost any given amino acid. For example, while leucine may be represented in the genetic code by any of six different codons (Table 10.2), structural genes in different species and even within different organelle genomes (e.g. nucleus and mitochondrion) display marked 'preferences' for particular codons (Grantham *et al.*, 1986).

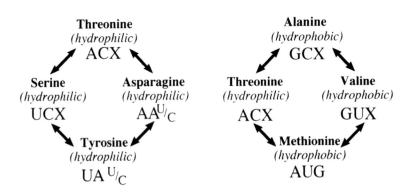

Figure 10.6 Two possible sets of evolutionary transition sequences between amino acids together with their corresponding codons and chemical properties. In the left-hand figure, both pathways between threonine and tyrosine pass through similarly hydrophilic intermediates. In the right-hand figure, the transition series from alanine to methionine requires two base changes but the route via threonine involves a change from hydrophobic to hydrophilic and back again and therefore is less likely to involve an acceptable intermediate protein.

There is still no proven causal explanation for this bias although there is evidence to suggest that it may be correlated with the availability of tRNA types. If this is so then it is likely that there may be a selective organism- or even organ-specific bias against certain codons. It is clear therefore that codon bias has the potential to alter at least to some extent the weighting of particular amino acid transitions that ought to be applied under the minimum nucleotide replacement scenario, but unfortunately this weighting system would probably need to be different for different taxa and different sorts of genes.

10.5.2 *Merits of minimum nucleotide replacement analysis*

An interesting discussion of analysing protein sequence data is provided by Fitch (1984) who considers the relative merits of conducting parsimony analysis directly on amino acid data (i.e. minimizing the number of amino acid replacements needed) as opposed to analysis based on minimum nucleotide replacement values. Fitch shows that by using the minimum nucleotide replacement values, one in effect increases the number of characters available for analysis. Further, in his example, based on bacterial cytochrome sequences, the distributions of possible tree lengths is considerably more skewed towards greater lengths in the case of the minimum nucleotide replacement analysis thus emphasizing the potential worth of the most parsimonious tree.

CHAPTER ELEVEN

NUCLEIC ACID METHODS

> . . . the potential molecular data set is incredibly extensive and, when
> fully utilized, should provide a detailed record of the history of life.
>
> Hillis (1987)

11.1   Nucleic acids in taxonomy

Direct consideration of the huge amount of phylogenetic information
comprising the base sequences of cellular nucleic acids is relatively
recent and expanding at such a prodigious rate that it is now within the
abilities of many laboratories to obtain informative sequence data from
a wide range of materials. However, the first direct taxonomic uses of
DNA information were far cruder than those presently available, and
included such techniques as assessment of the total nuclear DNA
content by means of **Feulgen staining** followed by microdensitometry,
and determinations of bulk nucleotide composition following acid
digestion of extracted DNA. Indeed the latter is still important in the
taxonomy and systematics of bacteria.

Recent advances in DNA technology have created a wealth of new
opportunities for taxonomy. On a daily basis, it is now making
available many more potentially informative taxonomic characters in
the form of nucleotide sequences than any other form of data,
although admittedly only a proportion of gene sequences are put
explicitly into taxonomic usage. When it became apparent that it
would soon be possible to sequence DNA from almost any organism
with comparative ease, many must have imagined that we would soon
know the true tree of life. However, because it is a new technology we
are still only beginning to understand its problems as well as its
potential.

11.2   Nucleic acids in cells

In prokaryotes all the genetic material is generally contained within a single circular double-stranded DNA molecule which is usually referred to as a chromosome. In eukaryotes the bulk of their genetic material is incorporated in somewhat more sophisticated chromosomes within the membrane-bound nucleus. Indeed, for many years after the discovery that DNA carries the genetic information it was thought that all the DNA in a eukaryotic cell was restricted to the nucleus. The first evidence to the contrary came from studies of **Feulgen**-stained preparations of certain plants and algae which showed a distinctly positive staining reaction localized within their chloroplasts. Following the initial discovery, reports quickly amassed to indicate that strands of DNA similar to those present in prokaryotes are present in both chloroplasts and mitochondria, referred to as cpDNA and mtDNA, respectively to distinguish them from the DNA contained within the cell nucleus. All organelle DNAs thus far isolated are circular molecules with the exceptions of those from the mitochondria of ciliate protozoans although early studies involving less careful extraction procedures frequently reported linear molecules. Because of their relative simplicity and ease of purification, organelle DNAs have been extremely important in unravelling phylogenies and therefore a special section below is devoted especially to both cpDNA and mtDNA.

11.2.1   *Nuclear DNA*

The nuclei of eukaryotic cells contain relatively enormous quantities of DNA. Eukaryotes typically have values ranging between 1000 and 10000 Mbp (megabase pairs). However, some have a great deal less. Horse-chestnut trees (*Aesculus hippocastanum*), the small and much-studied cruciferous plant, *Arabidopsis* and the nematode worm *Caenorhabditis elegans*, for example, all have genome sizes in the region of 100 Mbp while at the other extreme comes organisms such as some lilies with almost 100 000 Mbp of DNA in each nucleus. In all of these organisms, the number of functional genes and their associated regulatory and flanking sequences only account for about the same amount of DNA (approximately 100 Mbp) and so nearly all of the thousand-fold variation in the amount of nuclear DNA in eukaryotes is the result of non-coding or 'redundant' DNA.

Each functional gene, with a few important exceptions, is represented in the haploid genome by one or at most a few copies, although gene duplication has been a frequent occurrence in evolutionary history. In fact we probably ultimately owe the origins of most structural protein genes to this process as is evidenced by the recent finding that the genes of *Caenorhabditis elegans* fall into a relatively small number of families. A few functional genes are represented in the genomes of higher organisms by many copies, notable among which are the genes for ribosomal RNAs and for the various transfer RNAs.

11.2.2   *Repetitive DNA*

The nuclear genome of most eukaryotes not only contains the largely single copy DNA that comprises the functional genes and their various regulatory units, etc., but it also typically contains between 25% and 80% nonsense sequences of varying length which are repeated many dozens to many thousands of times. Short repeated sequences are usually arranged in tandem, that is end-to-end, in large blocks often near to the chromosomal centromere. Other repeated sequences which may have repeat units of hundreds or more base-pairs length, are called interspersed repeated sequences, which as their name implies are spread widely throughout the genome. In the case of one such group of sequences, the human Alu sequence family, each cell may contain as many as 300 000 and possibly up to 900 000 copies of the basic 300 base-pair sequence. Some highly repeated sequences often form a separate satellite band from the remainder of the DNA under gradient centrifugation conditions as a result of its fixed-base composition and consequently its distinct density. The formation of a satellite band gave rise to the common name, **satellite DNA**, which is often applied to highly repetitive DNA. It is also possible sometimes to isolate tandemly repeated sequences by digestion with an appropriate restriction endonuclease which, cleaving the DNA at the same relative position in each repeat produces a large number of copies of the repeat from every nuclear digest, and these can be separated electrophoretically. This technique has been used, for example, to investigate the origins of the Cetacea (whales and dolphins) where the discovery of the same highly repetitive component in both whalebone and toothed groups of whales has provided good evidence in support of monophyly of the order.

11.2.3 *Mitochondrial DNA*

In higher animals the circular, double-stranded mitochondrial DNA (mtDNA) molecule comprises approximately 16 000 base-pairs encoding two rRNAs, 22 tRNAs and usually 13 other genes mostly coding for proteins involved in the electron transport system located on the inner mitochondrial membrane (Moritz *et al.*, 1987). Unlike nuclear genes, animal mitochondrial genes lack **introns**, that is non-coding lengths of DNA inserted in and separating the coding parts of a gene into two or more parts. Further, most genes abut directly against one another or are only separated by short non-coding sequences. The genetic code employed in mitochondria is slightly different from that of nuclear genes. Although not exactly small, mtDNA molecules are relatively manageable and easily extracted making them one of the best studied pieces of the eukaryote genome and for a few organisms including man, mouse, cow and *Drosophila* the entire nucleotide sequence is known and additional complete sequences will no doubt follow in fairly rapid succession. However, most phylogenetic work on mtDNA to date has been based on either partial sequences or length variations found in fragments produced by restriction endonucleases (see section 11.5).

Compared with those of animals, the mitochondrial DNAs of fungi are far more variable although they encode for almost identical sets of proteins and ribosomal RNAs (Sederoff, 1984). In fungi, the circular mitochondrial DNA comprises some 18 to 100 kilobase pairs and often has intervening non-coding sequences between functional genes. In higher plants, mitochondrial DNA is almost mostly represented by a large circular molecule but sometimes smaller additional circular and perhaps linear molecules may also be present (for example in *Brassica*). The total size of the mitochondrial genome in higher plants (angiosperms) ranges from 100 to 240 kilobase pairs and is therefore considerably more difficult to work with than the smaller animal mtDNA. Some protista such as various species of *Paramecium* and *Chlamydomonas* seem to have linear mtDNAs.

Mitochondria are present in all eukaryotes except for four phyla of Protista and a few fungi (Cavalier-Smith, 1987). In all phyla mtDNAs show many similarities in the set of genes that they encode. Nevertheless, mtDNA in at least some groups evolves at a relatively high rate. Within the primates and other mammals, for example, mutations accumulate at a rate 5 to 10 times greater in mtDNAs than in single copy nuclear genes, although rates of change in nuclear and mitochondrial genes are much more similar in the echinoderms (e.g. sea urchins) and some insects. It is not yet absolutely clear whether the discrepancy in evolutionary rates between mtDNA and single copy nuclear DNA in mammals is due to a speeding-up of mtDNA evolution or a slowing down of change in the nuclear genome.

Most of the sequence differences observed in mtDNAs of closely related species are either selectively neutral silent changes at 3rd codon positions or transitions in ribosomal genes. Indeed, among various human populations it has been shown that transitions are approximately 24 times more common than transversions. The high rates of change found in many mtDNAs means that populations will often be polymorphic for particular nucleotide substitutions as there will have been insufficient time for most changes to become fixed in the population. Many mtDNA sequences have therefore been extremely useful in analyses of relationships between populations within species and between closely related species although some sequences are more conserved. The order of genes around the mtDNA molecule is rather more constant but differences are apparent between vertebrates, insects, nematodes, etc. and analysis of the gene sequence may prove to be a powerful tool in unravelling relationships between major groups.

One special feature of mitochondrial and of chloroplast DNA at least in most eukaryotes is that it is inherited only through the female (macrogametic) line.* In animals and plants the mitochondria in sperm or pollen grains are highly modified and on fertilization are lost. An important consequence of this maternal transmission is that differences that accrue in mtDNA are due entirely to mutation and are not the result of independent assortment or recombination. Further, because the mtDNA of a higher organism can only have come from its 'mother', whereas the nuclear DNA will contain a mixture of genes from both parents, a phylogenetic comparison of both has the potential to reveal past instances of hybridization. For example, using this argument, Song and Osborn (1992) were able to demonstrate that oil seed rape, *Brassica napus*, is not only a hybrid but one that has multiple origins. Further, lack of variation in mtDNA within a species

---

* The universality of this statement has recently been disproven for some marine molluscs (Zouros *et al.*, 1992) and it remains to be seen whether paternal inheritance of mtDNA is far more widespread than previously expected.

can be indicative that the species has in the recent past gone through a population bottleneck in which much of the genetic variation has been lost. Such is apparently the case with the cheetah, *Acinonyx jubatus*, and possibly also for humans, though the suggestion that all humans are the descendants of a single, prehistoric 'Eve' living some 200 000 years ago based on Cann *et al.* (1987) analysis of restriction fragments of mitochondrial DNA, is now widely rejected. The case of the data of Cann *et al.* on restriction site distributions of human populations is particularly revealing as it emphasizes that analysis procedures are crucial for correct phylogenetic inference (Gibbons, 1992).

Comparison of mtDNA nucleotide sequences, especially the genes for mitochondrial ribosomal RNA, have shown far greater similarities with certain prokaryote genes than with eukaryotic nuclear ribosomal RNA genes, an observation that led to the endosymbiotic theory of eukaryote evolution (Margulis, 1988). A further feature of interest is that there is some evidence that genes from chloroplast DNA may have become transferred to the mitochondrial genomes in some plants (Sederoff, 1984).

Purification of mtDNA is often fairly easy since for many tissues careful homogenization followed by differential centrifugation can yield reasonably pure samples of intact mitochondria free from cell nuclei. Subsequent density gradient centrifugation of extracted DNA can be used to remove any contaminating nuclear DNA fragments. Because most cells contain many mitochondria even small tissue samples can yield significant quantities of mtDNA. For example, 99% of the total cellular DNA in toad eggs is contained within their approximately 10 million mitochondria although at the opposite extreme come some yeast cells which may contain fewer than 10 mitochondria and some cultured mouse cell lines whose hundred or so mitochondria account for less than 1% of their total DNA (Borst *et al.*, 1983). However, nowadays PCR has largely eliminated the need for such procedures in taxonomic work.

11.2.4 *Chloroplast DNA*

As with mitochondria, chloroplasts also have their own circular DNA molecules, referred to as **cpDNA**, which are generally between 135 000 and 160 000 base pairs long and code for chloroplast ribosomal RNAs as well as a variety of other proteins (Palmer, 1985; Crawford, 1990). The complete gene sequences for several cpDNAs are now known and partial sequences are available for many others. Compared with

mitochondrial genomes, those of chloroplasts are relatively conservative and therefore their taxonomic application has tended to be at
species, genera and family level.

As with mitochondria, chloroplasts and their genomes are only
inherited through the maternal line and so like mtDNA, comparison of
cpDNA with the nuclear genome offers a potentially powerful tool for
revealing past hybridization events or introgression. Further, as
demonstrated by restriction fragment analysis (see below) of cpDNA it
can distinguish between potentially single and multiple origins of
hybrid taxa (Ogihara and Tsunewaki, 1988; Sytsma, 1990). Multiple
origins of the hybrid species, *Aegilops triuncalis*, for example, are
indicated by the presence in this taxon of two quite different
chloroplast genomes indicative that *A. triuncalis* has arisen through at
least two hybridization events, one with *A. caudata* as the female
parent and one with *A. umbellulata*.

11.2.5   *Ribosomal RNA and ribosomal genes*

Ribosomes are essential for the production of proteins in all living
organisms and consequently evolved very early on and are ubiquitous
among living things. Not only that, but they are abundant in that
virtually every cell has a large number of ribosomes and in the vast
majority of eukaryotes, their constituent RNA and proteins are
encoded by multiple gene copies making them particularly amenable
for sequencing. Again, while it is not surprising given their age that
ribosomes show some variation between organisms, they are on the
whole remarkably similar in both gross structure and in sequences even
in the most disparate members of the living world.

All ribosomes comprise two subunits, the large and the small subunit
(LSU and SSU, respectively), each in turn comprising a sizable RNA
molecule and a number of protein molecules. The sizes of the RNA
molecules in each unit vary between organisms and between nuclear
and organelle ribosomes in a given eukaryote (Table 11.1). In
addition, ribosomes of some organisms have one or two smaller RNA
molecules. The use of the latter in phylogenetics is limited only by
their size and therefore their potential information content.

rRNAs have had a considerable impact on molecular taxonomy
because within each molecule there are domains with different average
rates of nucleotide substitution. Consequently, each subunit contains

Table 11.1 Sizes of major ribosomal subunits from different sources

Source	Small subunit		Large subunit	
	Sedimentation rate (S)	No. nucleotides	Sedimentation rate (S)	No. nucleotides
Prokaryotes	16	1500	23	2900
Eukaryote nuclear-coded	18	1800	28	4000
Vertebrate mitochondria	12	900	16	1500

some sequence information relevant to divergence in the distant past of say the Cambrian, as well as other more rapidly evolving sequences carrying information relevant to separations almost up to the present.

As techniques advance, it is to be expected that sequencing whole subunits will become commonplace. However, at present it is usual to restrict sequencing to selected segments of one or other subunit selected because of their average substitution rates. This selective sequencing is achieved, for example, by use of the **polymerase chain reaction** (see section 11.3.2) in conjunction with carefully chosen primer sequences. This is made possible by the fact that some regions evolve at such a low rate that their sequences will be identical or very similar across whole phyla. Thus once these relatively stable flanking sequences have been determined for one organism, they can act as primers (usually called **universal primers**) for many other organisms. Without doubt this has greatly enhanced the usefulness of rRNA sequence data at all levels of phylogenetic analysis. The 16S rRNA molecule (SSU), for example, has gained considerable importance in recent years in the unravelling of prokaryote phylogenies and this role is likely to continue expanding along with the body of available sequences for many other groups.

11.2.6   Transfer RNAs and the genetic code

Transfer RNAs (tRNAs) are, like ribosomes, ubiquitous in living organisms and form an integral part of the protein synthesis machinery in that they donate particular amino acids to the peptide chain being constructed by a ribosome. There are about 60 possible tRNAs coding for the 20 naturally occurring amino acids. In general they have been highly conserved and in addition to having the appropriate anticodon recognition site which is matched to the corresponding triplet code of the messenger RNA, they also all have a typical 'clover leaf' configuration.

Within prokaryotes, the total number of tRNA genes per cell is approximately 60, i.e. approximately one gene per tRNA species. In contrast, in eukaryotes tRNAs are usually coded for by multiple gene copies, up to a total of 8000 or so in the African clawed toad, *Xenopus*, for example. The reason for such large numbers of copies of tRNA genes is likely to be that cells probably have a large tRNA turnover necessitating a high rate of production. Indeed, the total cellular tRNA content at any one time can be about 20% of the total cellular RNA (Clarkson, 1983).

The genetic code which translates 61 different triplet codes into just 20 different amino acids is virtually invariant among living organisms, although the genetic codes of several ciliate protists are known to deviate quite markedly from the 'standard' codes (see section 10.5.1). Further, there are several distinct differences in the codes used by mitochondrial and chloroplast tRNAs compared with the nuclear codes (Grantham *et al.*, 1986). Unfortunately there have been few comparative studies of codon usage between higher taxa but as the pattern of codon usage is a genetically controlled character (by definition) it might throw considerable new light on phylogenetics at a range of levels.

11.2.7 *Prokaryote and viral genomes*

In prokaryotes the genome comprises a closed loop of double stranded DNA of varying length, usually referred to as a chromosome. Unlike that in the nuclear genomes of eukaryotes, the DNA of prokaryotes is not complexed with histone proteins. In addition to their main (nuclear) chromosome, many prokaryotes also have one or more extra pieces of DNA called **plasmids**. These do not code for any major metabolic or structural genes but can be involved in antibiotic resistance and sex (genetic exchange). Plasmids are replicated and inherited along with the main chromosome. These structures have been important in gene cloning (section 11.3.1).

In contrast to the relative uniformity of prokaryote genomes, those of viruses collectively show enormous variation. Some are DNA based as with higher organisms while others have an RNA genome. Although small, considerable interest has been focused on what viral genomes can tell us about the origins of viruses themselves, and especially whether viruses are the result of a single evolutionary event or whether they have had multiple origins from their various host groups. The latter scenario seems to account for perhaps the majority of viruses although some virus families are known to attack members of more than one kingdom.

11.3   Amplifying DNA and dealing with small samples

Two DNA techniques have been particularly important in allowing nucleotide sequences to be determined for large numbers of genes over recent years. These are **cloning** and the **polymerase chain reaction** (**PCR**), both of which enable the number of copies of a given length of DNA to be increased to such an extent that there is enough material available for subsequent sequence or restriction fragment analysis (see section 11.5). It is not intended here to describe these in any great detail and only a brief outline of the processes involved is provided. Anyone wishing to read about these in more detail is recommended to consult Brown's (1990) summary or the more detailed practical descriptions provided by Hillis and Moritz (1990).

11.3.1   *Cloning*

Cloning is a procedure in which fragments of DNA from a donor organism or specimen are inserted into a bacterium (*Escherichia coli*) where they are replicated until a large number of copies have been made. These can then be purified and sequenced.

In order to clone an organism's DNA to produce a sequence library, the original sample of double-stranded DNA is partially cut using a restriction endonuclease such as Sau3AI so as to produce a wide range of fragment sizes. The enzyme Sau3I is used because the fragments produced have one chain longer than the other, i.e. overhanging, and this is used in the next step which is to insert the cut pieces of DNA within either a **phage**, **cosmid** or **plasmid** which had also been cleaved by another restriction enzyme, BamH1. This enzyme likewise leaves an overhanging length of single-stranded DNA, but one which is complementary to that left by Sau3I such that when the sample fragments (**donor** DNA) are mixed with the cut phage or other **vector** DNA, the complementary overlapping ends will **anneal**. The sample DNA fragments are then covalently bonded in position in the vector DNA using the enzyme DNA ligase. Depending on the choice of vector it is possible to clone DNA lengths up to about 40 kilobases long.

Vectors with their inserted DNA fragments are then used to infect the bacterium, *E. coli*. Each infected bacterium will contain different pieces of the DNA of the donor organism and when these have been plated out on agar growth medium, these fragments will be replicated many times within each of the growing bacterial colonies, i.e. the donor DNA fragments will have been **cloned**. The next step is to identify which *E. coli* colony includes the gene that it is intended to sequence. This is usually done by hybridizing a short, radio-labelled segment of the gene (a probe) with a sample of each colony. The probe will only bind to colonies in which the corresponding sequence occurs and these colonies will then be grown on prior to extraction of the cloned DNA.

11.3.2 *Polymerase chain reaction*

Frequently, the taxonomist is in the unfortunate position of having only tiny amounts of material available for chemical studies. Of course, chemical techniques are getting better all the time enabling workers to cope with and sequence ever smaller quantities of nucleic acids and proteins. However, the fact that cells contain elaborate enzyme systems for faithfully replicating DNA has made this molecule particularly amenable for the investigation of microscopically small samples.

The key to an elegant procedure for amplifying DNA sequences lies in the existence of DNA polymerase enzymes which will, under appropriate conditions, synthesize the complementary strand to a single-stranded DNA molecule (Mullis and Faloona, 1987). This reaction can be performed readily *in vitro* and if conditions are optimized, new complementary DNA strands can be synthesized with very low error rates. If then this newly synthesized complementary strand is used as a template for another DNA polymerase, the result will be two identical copies of the original single-stranded sequence (Figure 11.1). Fortunately, this process can be carried out time after time, in each case resulting in a doubling of the amount of DNA, so that after a few cycles enough DNA could theoretically be synthesized from a single starter molecule to permit direct sequencing. This process is known as the **polymerase chain reaction**, or **PCR**, and its uses are not restricted to taxonomy but also include important (if currently controversial) roles in forensic medicine. Unfortunately, PCR is not entirely without problems and like any technique that deals with very small samples, it suffers from signal-to-noise type problems. In handling the tiny amounts of source materials available to forensic scientists, perhaps just a few hairs or a splash of blood, enormous care has to be exercised to minimize the risk of contamination, for example from the workers themselves, otherwise the result could be amplification of quite the wrong pieces of DNA with potentially awful consequences.

Not surprisingly, PCR is becoming an immensely powerful tool for the molecular taxonomist, and in an impressive example Smith *et al.* (1991) have shown how it can be used not only to enable a new rare species of bird, an East African shrike, to be recognized as a new species but has also formed the basis of its formal description while allowing the only known (and living) specimen to be returned to the wild minus just a small blood sample and a few feathers (see section 6.4.4).

Another area where PCR will undoubtedly have a major influence is in the molecular taxonomy of small organisms such as Protista which even from cultures are unlikely to yield enough DNA for bulk preparation without an excessive amount of endeavour. However, while PCR is important when it comes to dealing with small samples, it is also of enormous value because it allows amplification of particular sequences without the need for cloning.

A recent and exciting development involving PCR is a process referred to as **random amplification of polymorphic DNA**, or **RAPD**, in which random primer sequences are employed to amplify sections of genomic DNA of unknown function (Hadrys *et al.*, 1992). RAPD utilizes oligonucleotides as primers but, unlike PCR which is aimed at amplifying a specific gene sequence, only one primer is added to the

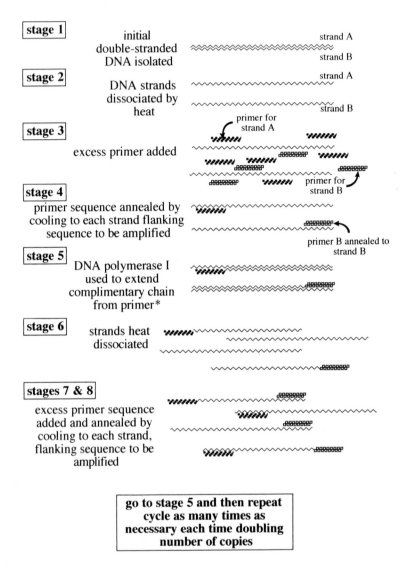

Figure 11.1 Diagram of processes involved in the polymerase chain reaction (PCR).

reaction mixture and it is a matter of chance how many sufficiently short sequences are bordered at either end by an inverted pair of that particular primer sequence so that they and the intervening sequence can be amplified. Typically an RAPD amplification using a 9 or 10 base primer may yield from 1 to 20 or so distinct amplified DNA sequences that can then be separated electrophoretically. Each band position appears to represent a true genetic entity that follows the rules of Mendelian inheritance, and thus can be regarded as a valid taxonomic character. Hadrys et al. (1992) provides a short review of the uses of RAPD which range from assessing kinship relations to the production of species-specific probes, assessment of whether there is gene flow between two samples and the detection of hybridization. To this may be added the unravelling of relationships, mostly between closely related species. Chapco et al. (1992), in a study of grasshopper RAPD, showed that approximately 50% of amplified DNA sequences were constant within a species while about 35% were constant (indistinguishable) between relatively closely related species. Greater phylogenetic information can simply be obtained by repeating the process with different random primer sequences. The attraction of RAPD analysis over traditional PCR is its ease and relatively low cost.

11.4 G+C content

In double-stranded DNA, each adenine (A) is hydrogen bonded to a thymine (T) and each guanine (G) to a cytosine (C). It follows that there must be an equal molar quantity of A and T and of G and C. However, there is no such restriction on the proportion of G+C to A+T and the ratio between these varies widely between different major groups of organisms. More importantly from a taxonomic point of view, within the prokaryotes there is exceptional variation with the molar percentage of G+C ranging from approximately 25% to over 75% in different taxa, although within any one genus the range is far less, typically only 10% to 15%. These differences even apply fairly well to fragmented prokaryote genomes suggesting that the differences may represent significant features of the whole genome rather than being due to the presence or absence of one or a few G+C-rich non-coding regions. G+C content has therefore been used extensively in bacterial systematics and has yielded a good deal of new information on the systematic relationships between a number of groups.

G+C content can be determined by a number of methods. Firstly, as three hydrogen bonds link G to C compared with only two between A and T, G+C-rich double-stranded DNA will stay associated to a higher temperature when the DNA is heated, i.e. it has a higher 'melting' temperature. DNA strand separation is associated with a marked increase in absorbance of UV light at 260 nm thus affording an easy way of quantifying the degree of melting. Other, less efficient means of studying G+C content that have been used in the past are chemical analysis after hydrolysis and CsCl gradient centrifugation.

11.5   Restriction fragment analysis

Many enzymes which are involved in the processing of nucleic acids commence their actions only at sites with a specific nucleotide sequence. Among these are a class of enzymes which cut DNA molecules within their own particular recognition sequences, and these are called **restriction endonucleases** or more simply **restriction enzymes**. For example, the commonly used restriction enzyme known as BamH1 cuts between the first two Gs of its recognition sequence GGATCC, while the enzyme HindIII cuts between the first two As of the complementary recognition sequence AAGCTT. The significance of restriction enzymes for phylogenetic research lies in the fact that in many organisms the distribution of sites in the genome where a given enzyme can cut is relatively fixed showing variations generally between species or genera.

Briefly the procedure used for **restriction fragment length polymorphism (RFLP)** analysis depends on the length of the DNA sequence and its purity. Restriction enzymes differ in the length of the sequence of bases that they recognize and cut at; 4, 6 and 8 are quite common. As each nucleotide position in a DNA molecule can take any one of four values (A, C, G or T), a restriction enzyme with a four base recognition sequence will on average cleave a DNA molecule every 4^4 bases (i.e. every 256 bases) while with a six base recognition sequence the average resultant length of DNA fragments will be 4096 bases. These fragments can then be separated electrophoretically as shown in Figure 11.2. Direct examination of restriction fragments is possible

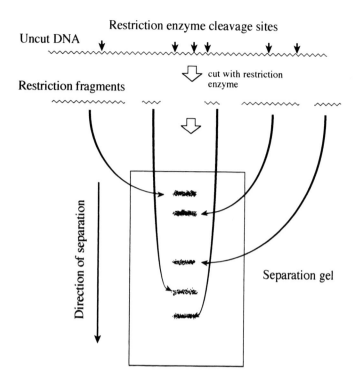

Figure 11.2 Diagram illustrating processes involved in restriction fragment analysis. Fragments are usually visualized on the separation gel by hybridization with a radio-labelled probe but direct visualization may be possible if a large quantity of a pure DNA sample containing a very large number of repeated sequences is involved.

with organelle DNAs which can be isolated to high purity and are small enough to yield relatively few restriction fragments. For example, application of a six base-recognizing enzyme to avian mtDNA will usually produce between 1 and 6 fragments while a five base-recognising one will produce between 2 and 9 sites (Shields and Helm-Bychowski, 1988). However, even with relatively small bacterial genomes, any given restriction enzyme will produce a vast number of different DNA fragments which when separated electrophoretically will yield little more than a smear. Therefore, in order to make information on the positions of restriction sites available from their electrophoresis pattern, radioactive probes are used to label particular fragments containing the complementary DNA sequence to that of the probe as illustrated in Figure 11.3.

Different restriction enzymes, with their different recognition sequences, cleave DNA at different places. Further, for each restriction enzyme it is possible to use a large number of different probes, each of which again recognises a different sequence. Using a range of combinations of restriction enzyme and probes it is therefore possible to obtain a great deal of phylogenetic information on the presence or absence of restriction sites. Even so, because of the large numbers of restriction fragments that might be revealed using a large nuclear genome, restriction fragment length polymorphism (RFLP) analysis is most easily employed to examine the relatively tiny amounts of DNA in organelle genomes.

Application of a single restriction enzyme can only provide information about the distances between restriction sites, combining data from two restriction enzymes applied separately, but in a mixture can provide additional information in the form of the sequence of the restriction sites within the circular genomes of prokaryotes or organelles. This data can form the basis for producing organelle genome maps and can reveal sequence rearrangements that may have considerable phylogenetic significance.

An important assumption of most RFLP analyses is that most of the

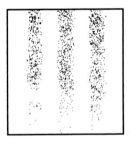
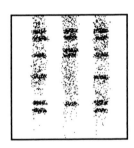

Restriction
fragments separated
by electrophoresis

Probes hybridized with
restriction fragments and
autoradiographed or
fluorescence imaged

Figure 11.3 Restriction fragments after cleavage of nuclear DNA results in a large number of different sized fragments which after electrophoresis would give a smear on the gel. Particular restriction fragments are identified by hybridizing them with a probe which only binds to those fragments with the complementary nucleotide sequence.

observed variants will be due to the gain and loss of restriction enzyme recognition sites rather than due to insertions or deletions of small lengths of DNA between two restriction sites, an assumption that appears to hold in the majority of cases (Shields and Helm-Bychowski, 1988). A fairly detailed appraisal of analytical and interpretational aspects of RFLP data is provided by Gillet (1991) who also describes the combination of restriction site analysis with the polymerase chain reaction (see below) as a way of selecting genome areas to be analysed and obtaining large enough quantities of restriction fragments to be visualized without the need for radio-labelled probes. Restriction site data are normally analysed from a cladistic point of view either using parsimony analysis or more recently using a maximum likelihood approach (DeBry and Slade, 1985).

Restriction site analysis has been used to trace events on a variety of evolutionary time scales from the origins of human races (less than 1 mya) to the origins of plant and animal families (of the order of tens of millions of years). The effective range is determined by the relative evolutionary rates of the genomic regions investigated. Thus, restriction fragment analyses of mtDNA are often used to help unravel evolutionary events during the comparatively recent past (at least in mammals) including intraspecific events such as the origins of human races. In contrast, restriction analyses of nuclear rRNA genes can provide evidence of events over far greater time scales. Overall, restriction site analysis can provide evidence on almost as wide a wide range of problems as DNA sequencing can, and is often more cost effective (Hillis and Moritz, 1990).

An important group of procedures based on restriction fragment analysis are various **DNA fingerprinting** techniques (Jeffreys *et al.*, 1985, 1990, 1991). These are based on the detection of the lengths of **minisatellite DNA** sequences that are distributed among animal (and probably plant) genomes. Any one 'family' of minisatellite DNA will comprise various lengths of a tandemly repeated DNA sequence distributed across many separate loci throughout the genome. Each repeat unit typically might comprise 30 or so base pairs within which a subset (core sequence) will be highly conserved across loci and between individuals. It is this core sequence that enables restriction fragments from many loci to be probed simultaneously after treatment of nuclear DNA with a single restriction enzyme. Thus a single restriction enzyme cut, probed with radio-labelled core sequence (or a tandem repeat including multiple copies of the core sequence), will reveal restriction fragments that may include a considerable number of minisatellite loci. Probably because of errors occurring during recombination, the number of repeats present at any given minisatellite locus is usually highly polymorphic within the population and so unrelated individuals are extremely unlikely to yield identical restriction fragment patterns. For the most part therefore, 'DNA fingerprinting' has been used to detect relationships between closely related individuals and it has found an important role in forensic medicine due to its ability to confirm identities from only tiny samples of tissues, etc.

11.6   DNA hybridization

When double-stranded DNA is heated to 100°C in aqueous solution the hydrogen bonds between the purine and pyrimidine nucleotides of the two strands break and the strands dissociate. On cooling homologous strands will reassociate again as the complementary base pairs realign. Similarly, once strands have reassociated, either to form **homoduplexes** (DNA all from one species) or **heteroduplexes** (comprising DNA from different species), the temperature at which they will separate again into single-stranded DNA, referred to as **melting**, is also dependent on the degree of similarity (serial homology) between the two hybridized strands. The process of allowing single-stranded DNA from different individuals or taxa to reassociate is referred to as **DNA hybridization**. By carefully controlling experimental conditions, melting or reassociation will occur over a fairly narrow temperature range and this can be used as a measure of the degree of sequence homology, and if the DNA sample contains a large number of sequences, the measure should in principle give a good estimate of overall genetic distance.

The key to being able to make use of information about dissociation temperature is the ability to separate double-stranded from single-stranded DNA and the most frequently employed method utilizes the fact that in phosphate buffer, only double-stranded and not single-stranded DNA will bind to hydroxyapatite. In practice, a hydroxyapatite column is loaded with a sample of double-stranded DNA at 60°C which then binds to the column. If the temperature of the column is raised, the double-stranded DNA melts and the resulting single-stranded DNA can be eluted from the column and measured (Figure

11.4). Quantifying the eluted DNA in hybridization experiments is usually achieved by having one of the DNA samples radioactively labelled and so the degree of melting can be determined by measuring the radioactivity of the eluent.

Before DNA hybridization can be carried out meaningfully it is first necessary to isolate those DNA sequences that are represented in the

a) experimental set-up

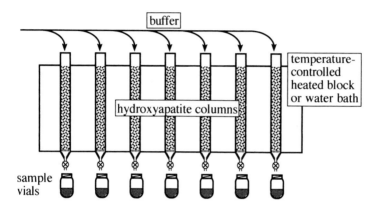

b) elution curve

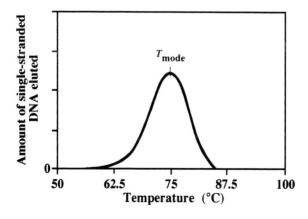

Figure 11.4 (a) Experimental set-up for DNA hybridization; (b) Stylized elution curve used for determining T_{mode} and for construction of cumulative retention (melting) curves (see Figure 11.5).

genome by only one or a few copies and so remove the highly repetitive DNA that would otherwise give misleading results. To do this, the DNA samples are first sheared to relatively short lengths (i.e. 300–1000 base pairs) typically by sonication. The sheared DNA is then dissociated by heating to 100°C followed by cooling to 60°C so that only the most homologous and readily-associating multiple-copy repetitive sequences will reassociate to form double helices which will bind to hydroxyapatite. The non-bound single-stranded single copy DNA is collected by washing it off the 60°C column.

In a hybridization experiment, a large excess of single copy DNA from one organism, the so-called **driver DNA**, is allowed to reassociate with a radioactively labelled sample from another, the so-called **tracer DNA**. Radio-labelling is relatively simple because strands of sonically sheared double-stranded DNA do not usually break at the same place thus giving rise to short single-stranded tails. DNA polymerase can then be used together with radio-labelled nucleotide triphosphates to complete the missing complementary pieces along the tail templates. Alternative radio-labelling procedures include nick translation and more recently, random primer labelling. Both techniques rely on the action of DNA polymerase which, on finding either breaks in one chain of a double-stranded DNA molecule (nicks) or randomly bound hexamers (or nonamers) on heat separated single-stranded DNA, forms a new complementary strand into which radio-labelled nucleotides can easily be incorporated.

11.6.1 *Interpretation of DNA hybridization data*

Having obtained melting curves for single copy DNA using (i) DNA driver and tracer from the same species (homoduplex reaction), and (ii) using tracer and driver from different species (heteroduplex reaction), a number of different measures can be applied to quantify the differences between the two curves. Three commonly used measures are the differences in median melting temperature (ΔT_m; Figure 11.5a), in the median melting temperatures corrected for different total amount of association (ΔT_{50H}; Figure 11.5b) and in modal melting temperature (ΔT_{mode}; Figure 11.4b) between the homoduplex and heteroduplex curves. In addition, the proportion of DNA that hybridizes in the heterologous reaction compared with the homologous reaction, referred to as the **normalized percentage hybridization** (**NPH**), is sometimes used. The smaller the values of

ΔT_{m}, ΔT_{50H} and ΔT_{mode} the greater the sequence homology between the samples. Measurement of ΔT_{50H} is fairly simple in that it is the temperature at which the greatest amount of DNA melting occurs and can be determined as the temperature at which the largest aliquot of eluted single-stranded DNA was obtained. The ΔT_{m} value is determined by plotting the cumulative amount of DNA eluted (i.e. melted) from each hydroxyapatite column (homoduplex and hetero-duplex reactions) against temperature with each one scaled from 0 to 100% eluted (or retained) as shown in Figure 11.5a. Then for each curve, the temperature at which 50% of the DNA has been eluted is termed the T_{m} and the difference between homoduplex and hetero-duplex values is the ΔT_{m}. However, this value is potentially misleading as the total amount of double-stranded DNA formed in the heteroduplex reaction is likely to be less than in either homoduplex reaction and this probably ought to be taken into account. This is achieved by relating the amount of DNA eluted from the heteroduplex column to the potential total amount as indicated by the homoduplex experiment. Having made this transformation, cumulative elution or retention curves are again plotted (Figure 11.5b) and the temperatures where the lines cross the 50% elution level are called the T_{50H} values.

Given the above three measures of melting temperature difference, the question arises which is the best one to use as a distance measure for the construction of phylogenetic trees. In other words, which gives the best estimate of sequence divergence? Sarich *et al.* (1989) have advocated that the best measure for phylogenetic inference is the ΔT_{mode}. This measure has the advantage that much empirical evidence suggests that it behaves in a more additive way than either ΔT_{m} or ΔT_{50H}.* However, a practical disadvantage is that accurate measurement becomes increasingly difficult as the DNAs diverge because melting of heterologous DNA tends to become spread over a wider range of temperatures. ΔT_{m} may be a reasonable measure for closely related species but has the disadvantage that it becomes increasingly insensitive as the amount of DNA that hybridizes in the heterologous reaction falls. Correction for the amount of hybridization in the heterologous reaction involved in calculation of ΔT_{50H} overcomes the insensitivity problem mentioned for ΔT_{m} and consequently ΔT_{50H} has

---

* DNA sequence data and measures relating directly to them cannot be truly additive because with increasing divergence, homoplaseous nucleotide substitutions will have a greater and greater distorting effect.

a) both curves normalized to 100%

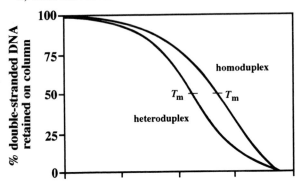

b) heteroduplex curve normalized to homoduplex curve

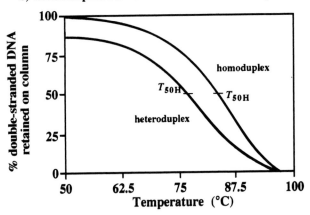

Figure 11.5 Melting curves for homo- and heteroduplex DNA 'hybrids' showing measurement of (a) T_m and (b) T_{50H}.

become the most widely employed measure for phylogenetic reconstruction. Importantly ΔT_{50H} has a fairly linear relationship with the degree of sequence divergence, but for small differences it may tend to overestimate distances and as with ΔT_m and ΔT_{mode} the accuracy with which it can be measured declines markedly if the amount of hybridization in the heterologous reaction falls below 50% of the homologous reaction value.

Ideally, for each pair of organisms, all four possible homo- and heteroduplex melting reactions should be conducted and average values of melting temperature differences used. Unfortunately the time and expense involved in radio-labelling samples usually means that only one of each is performed. Consequently, real hybridization data often comprise distances measured between a single species and all the rest.

As with distance-based immunological procedures, it is simple to show that to perform all pairwise comparisons between n taxa (ignoring replicates) would require

$$2 \sum_{i=1}^{n-1} i$$

or $n(n - 1)$ experiments. Another consideration relevant to DNA hybridization data is the accuracy with which parameters of melting can be measured and replication is highly desirable even though this will greatly increase the number of melting curves constructed.

Probably the largest and most extensive set of DNA hybridization studies conducted so far are those of Sibley and Ahlquist (1983, 1986, 1990) on birds. Avian systematics have for a long time been limited by a lack of new characters and an undoubtedly enormous amount of convergent evolution. For example, the 'vultures' of the New and Old Worlds represent a common **facies** that was quite independently derived from stork-like and eagle-like ancestors respectively, groups which may have diverged as long as 80 million years ago. Hybridization of bird DNA has been facilitated by the fact that avian red blood cells are nucleated and therefore quite small blood samples can readily yield a sufficiency of nucleic acid for this type of experiment, even making rare species potentially available for study since individuals need not be sacrificed to yield a sufficiently large DNA sample. However, while no one would dispute the magnitude of the task that Sibley and Ahlquist have tackled, their work is not without its criticism (see Cracraft, 1987) and consideration of arguments on both sides is informative about how hybridization data might be best presented. At this point however, it should be noted that Sibley and Ahlquist never claimed to have produced the last word on avian systematics and have positively encouraged others to carry out their own studies to improve on their ground-breaking work.

As Houde put it in 1987, 'the DNA hybridization technique has received much praise but little critical appraisal'. While admitting that DNA hybridization data may provide much valuable phylogenetic information, Houde emphasized the need for several potential problems to be addressed. In particular, for DNA hybridization data to act as a reliable measure of divergence, rates of DNA change should be uniform across lineages. This is usually assumed although there is ample evidence that at least with mitochondrial DNA rates may vary greatly from group to group. Sarich *et al.* (1989) expressed a similar concern that hybridization studies do not always satisfy the two criteria that are necessary before distance data can be used to construct phylogenies. These are that (i) reciprocal distances should be equivalent within experimental error, that is the distance between taxa A and B should be the same with either A or B used as the driver, and (ii) that the data must follow the triangle inequality, that is that if two taxa, A and B are more closely related to one another than either is to a third taxon, C, then distances A–C and B–C must both be greater than A–B. In an attempt to test whether rates of DNA divergence were consistent enough to be used as metrics for phylogeny reconstruction, Houde attempted to measure the consistency of distances between congeneric species and members of outgroup genera among primitive insect-eating passerine birds using some of Sibley and Ahlquist's earlier data, and found that these distances differed significantly between congeners. While this did not contradict the triangle inequality it did cast considerable doubt on the assumption of uniform rates of nucleotide substitution in different lineages. The significance of this to phylogenetic interpretation depends on just how large the disparities in evolutionary rates are compared with the separation between taxa. In conclusion, Houde postulated that while this potential source of error among the birds was probably not large enough to lead to the acceptance of wrong relationships, it is probably large enough to lead to ambiguity about relationships.

Another worry concerns the presentation of hard original data as mentioned by Harvey and Cotgreave (1991). There are several possible measurements that can be derived from DNA hybridization data and there is still considerable debate about the relative suitability of each of these as phylogenetic distance measures. Because of this, several authors have criticised workers who have chosen to present only one measure and/or not to present their original data (melting curves). Until analysis methods for hybridization data reach greater consensus, it is clearly desirable that all reasonable distances (together with any estimates of their variation) should be presented, even if only one is selected for tree construction at the time.

Detailed descriptions of DNA–DNA hybridization methodology are provided by Werman *et al.* (1990). Tree construction from distance data are overviewed in section 3.2.3.

11.7   Sequencing and associated methods

In recent years there has been a staggering number of advances in the field of molecular genetics making sequencing of DNA almost commonplace. Essentially DNA sequencing can be carried out in either of two ways. The DNA can be chemically broken down by specific chemicals as in the Maxam–Gilbert method or DNA polymerase can be used to build partial complementary strands from an existing template, the Sanger–Coulson method (Brown, 1990).

In the Sanger–Coulson method the Klenow fragment of DNA polymerase I is used to assemble a complementary DNA strand to that being sequenced. However, the strand so formed is caused to terminate prematurely after the polymerase enzyme has added a specifically modified nucleotide to the growing chain as illustrated in Figure 11.6. To start the reaction a sample of the particular single strand of the DNA to be sequenced is purified. Then four parallel-chain elongation reaction systems are set up, each containing the same complementary primer sequence which binds to part of the large DNA strand and initiates chain elongation by DNA polymerase. Each reaction also contains all four normal deoxynucleotide triphosphates needed for chain elongation, and in addition each mix contains a small amount of one of the four possible **dideoxynucleotide** triphosphates. Dideoxynucleotides are added to the complementary chain being built by the DNA polymerase just as ordinary nucleotides are, but having attached the modified nucleotide, no further chain elongation can take place. The particular modified nucleotide that caused termination of complementary strand synthesis therefore identifies the nucleotide that occupies the terminal position of the partially synthesized DNA strand. In practice one of the normal nucleotide triphosphates (typically adenine triphosphate) is radio-labelled (usually with ^{32}P) enabling the partially constructed chains to be located by autoradiography after

initial single-stranded DNA sequence
obtained by cloning or PCR

> 3'TTAAGACCTCAGCCGTATGCGGGTC . . . . 5'

primer sequence synthesized

> 5'AATTCTGGAGTC3'

primer sequence annealed to 3' end
of strand to be sequenced

> 3'TTAAGACCTCAGCCGTATGCGGGTC . . . . 5'
> | | | | | | | | | | |
> 5'AATTCTGGAGTC3'

complementary DNA sequences produced by DNA polymerase
I in presence of all 4 deoxynucleotide triphosphates (including a
radiolabelled one for subsequent localization of sequences) and a
small amount of dideoxy-GTP (G*)

primer-G*   chain extension
terminated after
incorporation of 1st G*

primer-GG*   chain extension
terminated after
incorporation of 2nd G*

primer-GGCATACG*   chain extension
terminated after
incorporation of 3rd G*

primer-GGCATACGCCCAG*   etc.

primer-GGCATACGCCCAG . . . .

Figure 11.6 Stages in carrying out DNA sequencing using the chain elongation blocking
properties of dideoxynucleotides (Sanger–Coulson method).

they have been sorted according to their size by electrophoresis
(Figure 11.7). Shorter chains migrate further in the gel and so the
constructed sequence is read from the end towards the base. By
adjusting the relative concentrations of normal and dideoxynucleotides,
the approximate range of lengths of the partially constructed strands
can be adjusted allowing a relatively long sequence to be constructed
by running a number of different ratios. Modern automated sequenc-

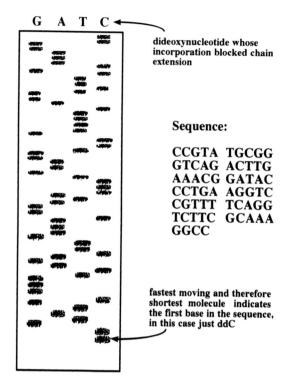

dideoxynucleotide whose
incorporation blocked chain
extension

Sequence:

CCGTA TGCGG
GTCAG ACTTG
AAACG GATAC
CCTGA AGGTC
CGTTT TCAGG
TCTTC GCAAA
GGCC

fastest moving and therefore
shortest molecule indicates
the first base in the sequence,
in this case just ddC

Figure 11.7 Representation of a DNA sequencing gel and corresponding base sequence.

ing machines use fluorescence rather than radioactive labelling techniques but the extra cost being counterbalanced by greater speed.

In the Maxam–Gilbert method, a purified sample of single-stranded DNA is first radio-labelled at one end and four aliquots of the DNA are each denatured by one of four chemical reagents that cleave at particular nucleotides. These are (i) dimethyl sulphate plus piperidine which cleaves at G, (ii) formic acid plus piperidine which cleaves at both A and G, (iii) hydrazine plus piperidine which cleaves at T and C, and (iv) hydrazine, NaCl and piperidine which cleaves only at C. Fragments produced by incomplete degradation are then separated by electrophoresis as in the Sanger–Coulson method and band positions noted. The difference in the banding pattern formed by reactions (i) and (ii) gives the locations of fragments terminating with A residues, while the difference between reactions (iii) and (iv) enables fragments terminating in T residues to be identified.

Nowadays, DNA sequencing is a routine business in many laboratories and a sequence once obtained is not generally publishable in its own right unless it can be shown to throw new light on a particular problem. In general, most new sequences are deposited in international computerized databases such as GENBANK or EMBL.

11.8 Conservation versus variability

The genome is no different from external morphology in that some features of it display wide variation even within species, whereas other parts may pass millions of years without accumulating many changes. Within the genes coding for almost any functional protein or functional RNA molecule there will be some regions that are functionally important. This can either be in an active way in that they code for reactive sites, or in a more passive way, for example, coding for amino acids that help maintain the tertiary structure of a protein through hydrophobic or hydrophilic interactions. In the cases of the relatively large ribosomal RNA molecules, this means that some quite long stretches have remained highly conserved over extremely long periods of evolutionary time while interspersed between these are regions that show far higher rates of nucleotide substitution. Thus a single rRNA sequence provides a series of data sets relevant to different taxonomic questions. By carefully choosing probes that recognize conserved sequences bordering a region that is generally known to show an evolutionary rate appropriate to the question under investigation, it is possible selectively to amplify and then sequence relevant rRNA domains.

11.9 Analysing sequence data

DNA or RNA sequences can provide extremely large data sets for analysis. Such data have been analysed in two essentially different ways, either using distance methods on reduced data, that is on taxa-by-taxa distance matrices derived from the comparison of the individual sequences (see section 3.2.3), or treating each variable nucleotide position as an individual character. The use of distance measures is largely historical in that computer programs and algorithms capable of dealing with parsimony or similar analysis of large numbers of characters have only been available in relatively recent years. Nevertheless, distance techniques are still widely used.

11.9.1 *Sequence alignment*

Having obtained base sequences from a number of organisms the next problem is to align them so that each base position can be compared across each taxon. As explained by Hillis *et al.* (1990), sequence alignment is a far from trivial problem because different interpretations of the correct alignment can have profound affects on the solutions obtained from cladistic analyses. Computer aligning can be very time consuming and the results very sensitive to the parameters chosen. Aligning by eye is often as good.

If the sequences have diverged only a small amount then sequence alignment is relatively simple, but with greater divergence comes greater homoplasy and increasing homoplasy makes alignment much harder. This problem is usually exacerbated further by the presence of insertions (additions) or deletions of various lengths. Indeed, two completely random sequences will on average show a 25% similarity. The process of obtaining the best alignment is itself a parsimony problem which attempts to maximize the number of bases that match in the two sequences being aligned while minimizing the number of insertion/deletion events. Numerous computer programs are available for sequence alignment, some better than others, and more are being produced at frequent intervals.

11.9.2 *Transition and transversion rates*

In double-stranded DNA, purine residues are always hydrogen bonded to pyrimidines, i.e. A to T and G to C. When mutations occur, they can be of two types, a purine or pyrimidine base may be replaced by the other purine or pyrimidine, respectively called a **transition mutation**, or a purine may be replaced by a pyrimidine or vice versa, called a **transversion mutation**. The significance of this to phylogenetic reconstruction is that transitions occur at a far greater rate than do transversions although as can be seen from Figure 11.8, there are in

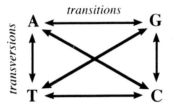

(i) there are twice as many
     ways that a transversion
     can occur than a transition

(ii) a transversion wipes out
     evidence of a previous
     transition but a transition
     does not wipe out evidence
     of a previous transversion

Figure 11.8 Annotated diagram showing two asymmetries between transitions and transversion substitutions. Note that despite the fact that there are more ways for transversions (A or G to T or C) to take place, transition substitutions are in practice far commoner, presumably because of chemical factors.

fact twice as many possible transversion mutation routes than there are transition ones. In conducting cladistic analyses of nucleic acid sequence data account should be taken of this difference, unfortunately just how much weighting would be justified is not so clear, and it is likely that there is no single weighting value that applies equally to all nucleotide positions within the genome. For example, in genes coding for proteins, there are likely to be differences in the rates of fixation of transversion mutations at each of the three positions in a codon. Transversions at the third position in the codons for 7 of the 20 amino acids will cause a change to an amino acid with a markedly different property while transitions only affect the amino acid in one instance (see Table 10.2).

In general, gene sequences of closely related taxa will differ predominantly in transitions because these occur at a far greater rate than do transversions. However, with increasing time since the taxa diverged, the numbers of transversions will also increase until for distantly related taxa the numbers of transition and transversion substitutions that appear to have taken place will approach a 1:1 ratio.

11.9.3   *Insertions and deletions*

If two DNA sequences differ in length by a single base pair then it is reasonable to assume that the difference has probably resulted from the single addition or deletion or a base pair in evolution, and such changes are best coded as a single character state change for the purpose of parsimony analysis. However, if the length differences are considerably longer than one base pair, then consideration must be given to how many separate instances of additions or deletions have taken place and this will usually depend on the distribution of apparent deletions/insertions along the length of the gene being sequenced. Thus if there appears to be a single insertion this would be best coded as a single character rather than a number of separate characters equivalent to the difference in sequence lengths.

11.9.4 *Paired and unpaired nucleotides in tRNAs and rRNAs*

Both tRNA and rRNA have secondary structures determined by the way the molecules fold back on themselves with pairing (hydrogen bonding) between complementary sequences of nucleotides on the same molecule. This intramolecular pairing has the consequence, with regard to using sequence data for phylogenetic purposes, that nucleotides taking part in pairing cannot change independently of their partners whereas unpaired nucleotides can. Therefore, if an unpaired nucleotide has no great functional significance it is more likely to change than a paired one and the latter will therefore be more conserved on average. Frequent insertions and deletions in these genes may make alignment a serious problem.

11.9.5 *A brief overview of the phylogenetic analysis of sequence data*

Parsimony analysis attempts to reconstruct phylogenies on the basis that evolutionary changes are rare events and therefore an acceptable evolutionary hypothesis should not require more instances of a rare event than is absolutely necessary (see chapters 1 and 3). Therefore the applicability of parsimony analysis to analysing DNA sequence data rests largely on whether the assumption that change is 'rare' is valid. Within a nucleotide sequence, each position has only four possible variants, and so if homologous sequences in different organisms have been subject to considerable evolutionary change it is likely that some positions will have changed in the same way in both daughter lineages giving rise to homoplasy. This is even more true as transition substitutions usually occur far more frequently than do transversions. It should be remembered that as taxa diverge and nucleotide changes accumulate, so too do reversals. When two sequences reach 75% divergence (i.e. they only have 25% of positions identical), they no longer carry any phylogenetic information and are effectively random. A corollary of this is that while differences between sequences from closely related species will tend to show an excess of transition substitutions over transversions, this proportion will decrease with increasing divergence between the taxa.

In parsimony analysis it is possible either to analyse all nucleotide positions together, or if the amount of change that this involves is considered too high, then the analysis can be restricted to the rarer transversion events (see section 3.1.3). In addition to maximum parsimony, several new methods are gaining widespread use. Of these the most significant are Lake's method of invariants and Nei's neighbour-joining method (Saito and Nei, 1987). These are described in more detail in section 3.1 and a thorough review is provided by Swofford and Olsen (1990).

Analysis of nucleotide sequence data can be confounded by a number of factors. Some organisms show a strong bias in the base composition of their DNA and if, for example, the DNA under investigation was very AT rich then transversions might be expected to be far more common than normal, and conversely, A to G or T to C transitions rarer. In this case it could be argued that transitions would be more informative which is contrary to normal experience. More problems occur in the analysis of protein-coding sequences because of selection pressures on the amino acids themselves. For this reason, it may at times be better to restrict analysis to bases at the third position of the triplet code that are largely redundant. Unfortunately, evolution at the third position is usually too rapid for this data to be very helpful in resolving other than close relationships.

11.10   Pros and cons of hybridization and sequencing

Different workers hold contrary views about which method provides the best phylogenetic evidence and probably there is no one correct answer as it will depend on the question being asked. In brief, the choice is between studying a small fragment of the total genome but in great detail and with many individual characters, or using hybridization to sample a far greater proportion of the genome but only obtaining distance measures whose accuracy is subject to experimental error. Another factor that has often to be considered is cost, and on this account DNA hybridization is often preferred.

Sequence information is particularly appealing from a cladistic point of view because it offers potentially vast numbers of discrete characters (base positions) that can each have one of four different states, i.e. A, T, G or C. In contrast, DNA hybridization only yields a distance measure and for that reason it has frequently been denigrated by proponents of sequencing. In a best case scenario, however, assuming hybridization values are adequately checked with reciprocal hybridizations and assuming that ample replications are carried out, inaccuracies in the hybridization method can be minimized to such an extent that the data effectively carry information about relationships that would otherwise require sequencing of some 10^5 bases (Felsenstein, 1987; Diamond, 1988). Given this it may seem strange that people bother to sequence DNA for phylogenetic analysis. However, while under ideal conditions it could be argued that hybridization is preferable, ideal conditions are hard to fulfil. First, they require a high level of technical skill and methodological consistency. Second, for hybridization results to fulfil the best criteria, measurements have to be made pairwise between all pairs of taxa (and reciprocally) which means that the number of experiments to be performed increases proportionately with the square of the number of taxa whose relationships are to be discerned, as explained in section 11.6.1.

11.11   Fossil DNA

The potential of the polymerase chain reaction described in section 11.3 extends considerably further than the analysis of DNA samples from living organisms. Its enormous power means that it even has the potential to make gene sequences available from at least some recently fossilized organisms. This possibility exists because DNA is a surprisingly robust and stable molecule. Thus, although fossilized material usually contains very little of its original DNA, and that which is present may be largely denatured, what small amounts there are can potentially be amplified many million-fold and sequenced as if it were new. Unfortunately, sequences retrieved from ancient specimens are typically very short, only a hundred or so base pairs long. Nevertheless, even with these and with carefully selected probes it should be possible to obtain phylogenetically usable data.

   Although efforts to date to extract DNA from mammoth remains have been largely unsuccessful there have been several notable success stories in this field. PCR has made it possible to obtain gene sequences from dried skins of the 13 000 year old extinct ground sloths from

South America and from the salted preserved skin of the recently extinct quagga, a relative of the horse and zebra from South Africa. However, these pale into insignificance compared with recently obtained results for 17 million year old magnolia leaves from lake deposits in Idaho (Golenberg *et al.*, 1990) and from stingless bees fossilized in Dominican amber approximately 55 million years ago (Cano *et al.*, 1992). Most fossil DNA is difficult to sequence because of oxidization of the bases, especially the pyrimidines, and the sequence-able fragments tend to be rather short. Surprisingly, the cretaceous bee DNA was extremely well preserved and lengths of some thousand or so base pairs could be amplified. A cautionary note should, however, be introduced as Lindahl (1993) casts some doubts on the validity of some of these claims. Clearly therefore, the circumstances of preservation are going to prove crucial to the success of obtaining usable DNA sequences from fossils and the best examples have been preserved under rather special circumstances. Thus in the case of the fossil magnolia leaves, they had become buried in a lake under volcanic clays which were deposited rapidly creating a virtually anoxic environment. Whether there could ever be any dinosaur sequences available for analysis still seems doubtful but there is every chance that DNA sequences will be amplified from skin and bone samples of many extinct organisms for which suitable material exists.

Such sequence information will no doubt help to resolve many phylogenetic questions relating to both recently and not so recently demised species. Whether attempts should ever be made to include recovered sequences into living relatives is open to debate.

CHAPTER TWELVE

PALAEOTAXONOMY, BIOGEOGRAPHY, EVOLUTION AND EXTINCTION

. . . the fossil record demonstrates, albeit incompletely, the actual course that evolution has taken. Until recently it was also taken for granted that fossils had a vital role to play in phylogenetic reconstructions . . .

(Hallam, 1988)

12.1 Palaeotaxonomy

Under appropriate circumstances, animals, plants and even microorganisms can leave remarkably good traces in rocks. In many cases the actual substance of the fossilized organism may have been replaced by minerals or at least have been greatly changed chemically but exceptions do occur, and fossils are sometimes formed under conditions that permit preservation of antigenic materials (Westbroek *et al.*, 1979) or even DNA that can be isolated and put to systematic use (Golenberg *et al.*, 1990). Some fossil taxa are described not from their own remains but from changes they have brought about around themselves when they were alive, for example, fossil animals can be described from their tracks, burrows or nests. Clearly, the organisms that created these **ichnofossils** will virtually never be known with any great certainty and so concepts of species or even genera and families in such cases may be very hazy. However, other better preserved taxa provide a historic record – albeit an incomplete one – of life on earth and thus provide a window into the past. Fossils therefore have a potentially important role to play in taxonomy and phylogenetics, but their interpretation and use is not without its pitfalls and contentious issues.

12.1.1 *Completeness of the fossil record*

Some organisms such as shelled molluscs and trilobites lend themselves well to fossilization due to their hard and durable exteriors. Needless to say, however, soft bodied creatures require special combinations of circumstances to permit their preservation, such as the rapid deposition of soft fine sediments and conditions of low oxygen so that bacterial degradation is slowed. Even so, some fine fossils of soft bodied organisms have been unearthed, including jelly fish. Thus, the type of habitat that an organism dies in profoundly affects its chances of fossilization. With few exceptions, such as the skeletons of larger animals, fossils of terrestrial animals are usually rather scarce. Even leaves are only preserved under rather special circumstances such as in the anoxic bogs that ultimately produced present day coal measures. Despite the vast numbers of terrestrial arthropods that have inhabited the earth for a million years, their fossils are seldom abundant and nearly always have resulted from the animal falling into mud or water or being entombed in plant resin later to form fossiliferous amber (Poinar, 1992; Cano et al., 1992). Suitable circumstances for fossilization of such organisms may indeed be very rare in both time and space, and if the chance of discovering such sites is added to this it is not surprising that serendipity has played a large part in our knowledge of many fossil groups. Furthermore, even if the members of a taxon existed in fairly constant numbers throughout a given period of geological history, circumstances of preservation and exposure of relevant fossil-bearing strata mean that their fossil record will be patchy. In reality, most taxa probably undergo times of abundance when the chance of fossilization is relatively good and periods of comparative scarcity when fossilization would be unlikely.

How little we know the histories of many groups can be well illustrated by two recent discoveries that push particular fossil records back by substantial amounts. Firstly there is the discovery of several well-preserved fossils of a species of tarantula spider (*Rosamygale grauvogeli*) which, at 240 million years old, pushes back the known ancestry of its group, the Mygalomorpha, by more than a factor of two (Selden and Gall, 1992). An even greater extension to a fossil record, in terms of number of years, comes with the discovery of specimens of the eukaryotic alga *Grypania* dated at 2100 million years old, that is some 300 million years older than the previous record for a fossil eukaryote (see Riding, 1992). Without doubt similar finds are certain to keep on occurring for many years to come.

12.1.2 *Interpretation of evidence*

Fossils are seldom if ever perfect. Fossilization almost invariably involves distortion, chemical degradation and replacement in addition to the normal processes of death and decay. Nevertheless, under fortuitous conditions even some very delicate organisms and fine details can sometimes be preserved in remarkably good condition. Unfortunately, the palaeontologist has to work on what material is available and this frequently means interpreting less than ideally preserved fragmentary or distorted material. Today, as in the past, the keen eye of the palaeontologist is usually of critical importance in interpreting what are often only faint differences in coloration or subtleties of topology. Thus, even with the best of intentions, the resulting interpreted figures risk showing bias and it is not unknown for past workers to have 'interpolated' missing structures to a lesser or greater extent. Advances in photography including the use of ultraviolet or polarized light may help a little in the presentation of results but they are no substitute for careful and critical examination of the original fossils.

Finally, we should also be aware of the possibility of deliberate forgery of fossil evidence for gain or otherwise. The Piltdown man saga shows how even experts can be taken in by a clever forgery while at the opposite end of the spectrum, two well-known astronomers, Hoyle and Wickramasinghe (1986), have claimed that the feather imprints of one of the world's most famous fossils, *Archaeopteryx*, are of similarly human origin. Fortunately for systematics, their view has not been widely accepted by other palaeontologists and Kemp's book review offers an alternative perspective.

12.1.3   *The palaeo-species concept*

Interpreting what is a species from fossils is not easy. Fossils are by their very nature incomplete representations of past organisms and often they are poorly preserved. As such, it is highly unlikely that they will display adequately the sorts of small differences that are so often used today to separate extant taxa although amber fossils might sometimes permit such studies (Poinar, 1992). Fossil 'species' will therefore only correspond to biological species notions if the latter happen to be well correlated with fairly obvious morphological differences. Thus if fossils are to be classified to species level, they may require a modified species concept. A corollary of the above is that we may expect that the numbers of species of a fossil genus will be underestimated both because many may not have been preserved or discovered and because 'true' species may not be distinguishable among the existing fossilized remains. Indeed, even among extant taxa we have little idea about how many apparent species are actually sibling species complexes although in some groups it may be quite a considerable proportion.

Two other perhaps more challenging problems relate to the fact that the full details of how species arise over time are not yet understood. First, Hennigian philosophy and modern cladistic techniques assume that all speciation events involve dichotomous (or in later models, polychotomous) splits (Figure 12.1 *right*). This contrasts markedly with the apparent change observed in many fossil records that suggests that speciation can occur over time without any necessity for splits to occur (Figure 12.1 *left*). Speciation which results from a lineage splitting into two (or more) separate species is referred to as **cladogenesis**, whilst speciation resulting from the accumulation of change over time in a single population is termed **anagenesis**.

Second, early Darwinian notions assume that evolutionary change is likely to be a gradual process with new species often forming over geologically active periods of time, so called **phyletic gradualism** (anagenesis). If this is the case then any near complete fossil sequence would be expected to show a gradually evolving sequence of forms and the taxonomist would be posed with the arbitrary problem of where to

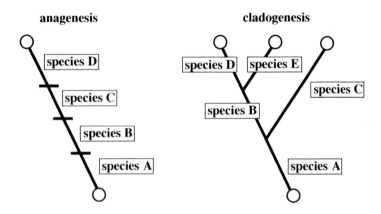

Figure 12.1 Two alternative modes of speciation, anagenesis in which morphological changes that ultimately define species boundaries accumulative over time, and cladogenesis in which new species arise after the development of some isolating factor. In cladogenesis it is generally but not always assumed that speciation events also mark the end of the ancestral species as illustrated.

draw the line between species. One major practical problem is that there are relatively few fossil series that are complete enough to permit assessment of phyletic gradualism. One such record is provided by the trilobites, a group of marine arthropods which tended to be well fossilized due to their hard exoskeletons. Thus Sheldon (1987) was able to examine some 15 000 trilobite fossils belonging to eight lineages and spanning an approximately 3 million year period from the Ordovician of Wales. His conclusion was that these eight lineages underwent gradual and parallel change in several morphological characters which in this case seems to support the idea of phyletic gradualism.

Evidence has also been accumulating to suggest that evolution is not always such a gradual and more or less continuous process, and instead that taxa may exist through prolonged periods during which little change occurs and that occasionally circumstances occur that give rise to brief periods of rapid evolution and speciation. This latter idea is now commonly referred to as **punctuated equilibrium** (Eldredge and Gould, 1972). Studies of several groups of Cretaceous dinosaurs have provided some interesting recent examples in which morphospecies appear to have remained constant for prolonged periods of up to 5 million years until their environment and/or population size was dramatically changed due to marine incursions after each of which a new morphospecies appears to have completely replaced their predecessors (Horner *et al.*, 1992). If this is a fairly general model of how evolution occurs it suggests that recognizing and naming palaeospecies may often be fairly easy, a matter of identifying morphological jumps between times of comparative stasis. Unfortunately, it is not yet certain which form of evolution is dominant and analysis of even some relatively complete fossil records have been used to argue in favour of both phyletic gradualism and punctuated equilibrium by different workers (Futuyma, 1986).

12.1.4   *Phylogenetic analysis and classification of fossil taxa*

Two separate issues have to be considered here, firstly what fossils can do to increase our understanding of the evolutionary history of organisms and how fossil taxa should be integrated into classifications of extant taxa, if indeed they should. The incompleteness of the fossil record is often quoted as an argument against including fossils in cladistic analyses but many data sets are incomplete and most extant species are still unknown to science. Perhaps therefore, the question should be whether the fossil record is sufficiently complete for the purpose at hand. The problem of incorporating fossils into the classification of living organisms does not so much reflect the quality of the fossil record but rather it concerns how a system designed to accommodate end points of evolution can be adapted to include the internal (ancestral) branches of the evolutionary tree.

12.1.5 *Phlogenetic aspects of fossil taxa*

Fossils provide two separate forms of phylogenetic evidence. At their best they provide additional taxa that can be scored for character states and included in a phylogenetic analysis, and they give a minimum estimate of the age of origin of particular taxa and their character states. Unfortunately, fossil taxa will seldom, if ever, be so well-preserved that they can be scored for all the characters that are available to the morphologist working with extant species so the data set they provide will almost inevitably be incomplete, in some cases very incomplete. Incomplete data sets tend to cause difficulties in phylogenetic analyses, at the very least giving rise to large numbers of equally parsimonious trees such that the consensus tree provides very little information. Patterson (1981) proposed that fossil taxa need not be considered in phylogenetic analysis. However, fossil taxa may show combinations of character states that are potentially very informative about relationships. Further, some taxa, commonly referred to as living fossils, may survive long after all their close relatives have become extinct, and no one would suggest that these should not be included in an analysis.

This problem has been examined in more detail by Donoghue *et al.* (1989) who compared cladograms obtained for amniote vertebrates and seed plants both including and excluding fossil evidence. Their conclusion was that incorporation of fossil taxa can have a substantial affect on the topology of the cladogram obtained and therefore should not be excluded. As pointed out by Novacek (1992) there is as yet only a comparatively small body of carefully analysed data that can throw a meaningful light on the importance of including fossils in phylogenetic analyses and therefore more work in this field is urgently needed. The main problem that is likely to be encountered is that inclusion of incomplete data sets for fossil taxa can greatly reduce the resolution of the consensus cladogram. The critical feature of a cladogram is its accuracy and a more resolved one should not be preferred if it is also likely to be a less good estimation of phylogeny.

12.1.6 *Inclusion in classification of extant organisms*

While some fossil taxa can clearly be placed in extant higher categories and some are clearly members of extinct higher groups, the more well known the fossil record becomes, the more likely it is that fossils will be found which actually represent a common ancestor of a group of living taxa. This poses the philosophical problem of how such fossil taxa should be classified. Suppose that a fossil is found which is indistinguishable from the hypothetical common ancestor of two extant named taxa. To which, if either, of these does the fossil taxon belong? Should a new equivalently ranked taxon be erected to receive the fossil taxon? Further, if any part of the fossil record was nearly complete, then how many hierarchic levels of taxa would be required in order to fit all the species in a Linnean classificatory system?

These problems have been considered at length by a number of workers (e.g. Crowson, 1970; Patterson and Rosen, 1977; Wiley, 1979) and have been reviewed by Wiley (1981). In brief, Wiley distinguishes three possible ways of treating fossil taxa. First, fossil taxa could be completely excluded from classifications of extant taxa and different classifications could be erected for the faunas of each historical epoch. This scheme could, however, mean that a taxon would be given different ranks in different epochs. At the other extreme fossil taxa could be included with extant taxa in the same classification and the two treated identically. This proposal has the drawback alluded to above that it is difficult to allow for true common ancestors (stem species) while retaining the idea of a phylogenetically compatible hierarchy of monophyletic taxa. One way around this is to treat all fossil taxa as if they are terminal taxa, although of course this may not be true. The third alternative is in effect a compromise in that it proposes that fossils be classified together with extant taxa but that they are covered by special provisions to overcome the difficulties they pose. This latter view is the one advocated by Patterson and Rosen and by Wiley. However, simply indicating extinct taxa with the traditional dagger mark (†) does not solve the problem of assigning the fossils to a higher taxon without creating large numbers of higher taxa with only one or a few included species. To overcome this it has been proposed that fossil species, appropriately indicated by a dagger, should be placed in a higher category specific to fossils, the **plesion**, which has no defined rank in the Linnaean hierarchy but which 'substitutes' for any higher category. The use of the plesion concept would not necessitate making any changes to the names of higher taxa that have already been proposed, as would be necessary if a taxon were for instance changed from superfamily to order status. This latter point is by no means trivial as with fossils the discovery of a new taxon of apparent intermediate status might ordinarily involve assigning a considerable number of taxa to new ranks.

12.1.7 *Stratigraphy, evolutionary rates and molecular clocks*

Fossils not only provide details of the structure and sometimes the biology of extinct organisms, they can also provide estimates of the ages of the groups concerned. Thus the stratigraphic age of a fossil also sets a minimum evolutionary age for any apomorphous character state that can be associated unambiguously with homologous structures in extant species. It has been argued that this information can be used to distinguish plesiomorphous from apomorphous character states although others have argued vehemently against this on the basis that the incompleteness of the fossil record does not permit exclusion of the possibility that a character state is derived even if it appears earlier on in the fossil record than another state.

With respect to the age of a taxonomic lineage, it should be noted that the literature is replete with examples of taxa, of all levels, whose fossil record keeps getting pushed back further and further, not just by small increments but often by considerable periods (see section 12.1.1). Dated fossils therefore usually only provide minimum ages for clades and if these are used to assess the rate of change of molecular sequences (protein or DNA), the continued discoveries of older representatives of any given taxon will mean the frequent and often substantial recalibrations of the molecular clock.

While different species or lineages undoubtedly undergo vastly different evolutionary rates, as evidenced by their external appearance, molecular studies have suggested that some DNA sequences (or those of the proteins they encode) may change with time at a much more uniform rate even across taxonomically different groups of organisms. If such sequences exist then the degree of divergence between the sequences in two taxa might correlate well enough with the actual time since the divergence of the two lineages that it could be calibrated and used to estimate the time since divergence of other lineages (Hallam, 1988). Calibration would of course depend on some groups that are particularly well represented in the fossil record such that the date of divergence of particular lineages could be estimated.

The **molecular clock** hypothesis has had a chequered history (Thorpe, 1982; Hallam, 1988) which will be briefly surveyed below. In general, molecular clock proposals fall into two types. Those concerned with a particular molecular sequence that is believed to have evolved in a sufficiently regular, clock-like fashion (e.g. glutamine synthetase, Pesole *et al.*, 1991) and those that average data from many genes, such as with genetic distances based on allozyme studies or even more extremely, on the whole genome as in DNA hybridization.

Certainly, genes that are subject to strong divergent selection pressures would be expected to make very poor candidates for molecular clocks because the amount of change they accumulate over time is likely to depend very much on the lineage in which they are found, and on the conditions to which they are exposed. Studies on genetic distances derived from enzyme polymorphism suggest that over a large number of taxonomic groups and enzyme systems studied, a Nei's genetic distance of 1.0 is roughly equivalent to a time' since divergence of 18–20 million years (Thorpe, 1983). Combining this with the average values of genetic distance found between congeneric species pairs (see Figure 10.5), Thorpe found that most species pairs diverged in the order of 3 million years ago and most genera might be about 23–30 million years old.

Molecular clocks making use of DNA–DNA hybridization studies at first sight seem attractive for deciphering phylogenies because the affinity values they yield integrate data from the whole genome. They might therefore be expected to overcome objections to molecular clocks based on single genes, many of which might show different evolutionary rates in different lineages as a result of natural selection.

In attempting to use molecular clocks to estimate divergence dates, another confounding factor is that different types of organisms may have very different average generation times. Since we will probably never know the actual generation times of most extinct organisms any estimates must be based on correlations with similar extant ones, thus creating yet another variable in the equation.

12.2 Biogeography

Biologically and historically the phylogenetic relationships between species and their geographic distributions are intimately linked. As far as is known, the majority of new species arise through allopatric speciation, that is a single species becomes segregated into two or more genetically isolated subunits by geographic barriers. These barriers may either form *de novo* such as with orogenic (mountain building) processes or species may become transported to new areas of their own accord and thus found new effectively isolated populations as has happened often on islands. At least 70% of species may have originated thus (Brooks *et al.*, 1992).

While biodiversity is concentrated around lower latitudes, the association between geographic isolation and the evolution of new taxa also has a profound effect on both the total diversity and the number of endemic species present in a particular region. Islands are well known for their high proportions of endemics while generally having far fewer species in total compared with adjacent larger land masses. From a conservation point of view this has serious consequences as the loss of an island's fauna or flora as a result of human activity can be expected to cause a disproportionate loss of biological diversity. Island biotas always have been vulnerable and over the long term most island species are doomed to extinction, but this does not excuse the tremendously accelerated extinction rates for island endemics that have happened during recent human history.

Disentangling the historic origins of the distributions of organisms is not a simple task, which is not surprising given the multiple ways in which distributions can arise (Croizat *et al.*, 1974; Ball, 1976). The dispersal characteristics of particular groups will have profound effects, for example large terrestrial mammals may be largely restricted to land connections in migrating from one region to another whereas corals or plants with aerially dispersed seeds may be able to propagate their species over vast distances with ease. Strategies for reconstructing historical events leading to present distributions will also depend on the ages of the taxa concerned. Thus for ancient lineages such as the reptiles, fossil evidence may be valuable in mapping historic ranges of taxa. In contrast, interpretation of recently isolated groups may be totally reliant for interpretation on phylogenies based on extant taxa.

With taxa that are not so diverged as to be reproductively isolated, phylogeny and distribution can potentially be confused by the retention of gene flow due to migration patterns between populations (Felsenstein, 1982b). Figure 12.2 illustrates the problem. The left-hand diagram shows a cladogram derived from analysis of the characters (for example gene frequencies and consequent genetic distances) of the five taxa A to E, and on the right-hand diagram shows migration patterns between populations and groups of populations of the same reproductively compatible taxa that, given an appropriate choice of migration rates along each route, will lead to the construction of the left-hand cladogram. If the geographic distributions of the five taxa seems to correspond to a migration model derived from a cladogram then it would be impossible to reject either hypothesis, migration effects or evolutionary relationship without migration, without additional information. On the other hand if the migration model does not, on geographic grounds, offer a plausible explanation, for example in Figure 12.2 if populations of A and E were far closer to one another than A was to B or E to D, then the phylogenetic hypothesis would seem to be far more likely than the migration one.

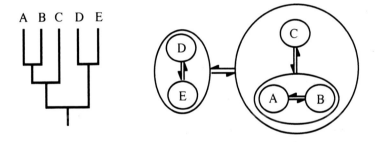

Figure 12.2 Diagram showing that any cladogram can in theory be produced by a pattern of migrations between potentially interbreeding populations (based on Felsenstein, 1982b).

12.3   Coevolution

Another key feature in understanding evolution and biodiversity, although a surprisingly poorly understood one, is the possible role played by coevolution. It is clear, for example, that the number of species is limited by the number of available niches, and therefore if the niches involve other organisms then correlations are to be expected between the numbers of niche-creating species and the number of different occupants of these niches. Taking again as an example insects that are parasites on other insects, the total number of parasitic species is likely to be limited in some way by the number of different types of hosts. Thus in habitats where there are many species of butterflies, there are likely to be many species of parasitic flies and wasps. However, the question arises how the niches, in this case potential host insects, come to be occupied. Several mechanisms, including classical coevolution, can usually be postulated in any case and the relative importance of these can potentially be determined using phylogenetic methods.

Detection of coevolution depends on the availability of cladograms for the two groups of organisms that are suspected of having coevolved, and also on a knowledge of the times at which the various lineages diverged. As illustrated by Mitter *et al.* (1991) two clades of organisms can become associated in a number of different ways. Species may be associated as a result of coevolution with splits in a lineage within one group being mirrored by equivalent splits in the corresponding members of the other group, with these splits occurring at the same time (Figure 12.3a). In this case the cladogenesis of one group leads to cladogenesis in the other group, for example if a host species became split by some vicariant event, then the populations of a parasite species would become simultaneously subdivided and each will continue in its own evolutionary arms race with its particular, now isolated and diverging, host. This sort of process will give rise to perfectly concordant cladograms between the two associated groups of organisms. Apart from this true association by descent, it is possible to recognize two other forms of association which do not reflect coevolution. First, a situation can arise which involves no evolutionary association with the result that the cladograms of the two groups will show no significant concordance (Figure 12.3b). Second, it is possible that phylogenies could display significant concordance but without this being the result of coevolution. In this latter case, it is not sufficient to know the evolutionary branching patterns of the two groups, but it is also necessary to have knowledge of the times at which various lineages split. In the case of coevolution, not only should branching patterns of the two cladograms be concordant but the dates of the splits would also show a high level of correspondence that (by definition?) would not be the case in the absence of coevolution.

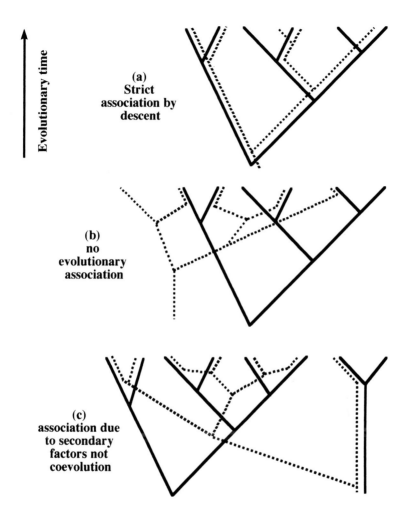

Figure 12.3 Three different possible associations between two groups of biologically related organisms (e.g. predators and their prey, models and their mimics, parasites and diseases and their hosts) (based on Mitter *et al.*, 1991).

12.4   Phylogenetic trees and the pattern of evolution

In this section some relatively recent work that uses information gathered from many different phylogenetic studies is considered to try and gain an insight into what patterns and evolutionary processes may actually follow and, at the risk of circularity, whether phylogenetic results fit current expectations of the patterns and processes of evolution.

Much of what has gone before has indicated that cladistics takes ideas about how evolution works and uses these to derive techniques for phylogenetic reconstruction. For example, maximum likelihood procedures for reconstructing phylogenies from DNA base sequences make specific use of models of how bases are believed to change in evolution, which in turn are largely derived from empirical studies of sequences. Equally, application of parsimony can be justified on theoretical and philosophical grounds providing that evolutionary changes are sufficiently improbable (Sober, 1988). However, if the biological assumptions implicit in a model of phylogenetic reconstruction are clearly understood, then the output of the model (in this instance cladograms) should in theory provide a way of testing whether evolution actually proceeds in the way the model assumes. For example, interest has recently been gathering about whether all taxa are equally likely to speciate or go extinct and whether evolution proceeds gradually or in jumps. The problem is how to carry out the comparison which seems to require knowledge of 'true' rather than estimated phylogenies.

One way of approaching this problem is to ask whether the distribution of phylogenetic trees obtained from cladistic analyses conforms to expectations based on different models of evolution. First, it is necessary to consider what expectations we should have of the likelihood of encountering different tree structures. In an early foray into a related topic, Rosen (1978) considered each possible tree topology to be equally likely. However, consideration of the ways in which different tree topologies can arise through the evolution of a new taxon from any of the terminal (i.e. extant) taxa of a smaller set of taxa will show that this is not the case. A simple rooted phylogenetic tree with two terminal taxa is shown at the top of Figure 12.4. If a third taxon were to evolve from either of these, the two resulting three-taxon trees would each have exactly the same topology although there would be two arrangements of the taxa involved. If a fourth taxon now evolves from one of these three taxa there are two possible resulting tree topologies (one symmetric, one asymmetric). However, these two topologies are not equally likely to occur because of the three possible taxa in a three-taxon tree that a fourth could evolve

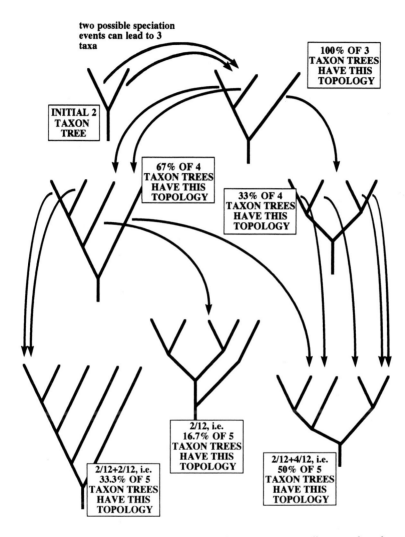

Figure 12.4 Construction of trees with progressively more taxa split at random from existing terminal taxa (thus resembling the process of random speciation) and the probabilities of particular unlabelled tree topologies being derived from the numbers of ways each tree topology can be derived through this branching sequence (Harding, 1971).

from, only one gives rise to the symmetric four-taxon tree whereas evolution from either of the other two taxa will give rise to topologically identical asymmetric ones. Immediately we see an inequality which continues through all further bifurcations of a terminal taxon. Using this protocol (Markov method) it is possible to calculate the probabilities of obtaining a tree of any given topology through the evolution of extra terminal taxa from existing terminal taxa.

Expected tree frequencies were recently compared with a large sample of 'real' tree topologies taken from published phylogenies by Guyer and Slowinski (1991) with surprising results. These workers showed that frequency distributions of tree topologies obtained by phylogenetic studies in three major groups of organisms do not uphold the expectations from random evolution of taxa. In fact, the observed distributions corresponded best to the idea that each distinct labelled tree rather than unlabelled tree topology is equally likely. The reason for this apparently unlikely result is not clear as there are two obvious possibilities. First, it may be that the evolution of new species does not occur randomly across all extant taxa and indeed this may not be unexpected as organisms might be expected sometimes to encounter new situations enabling one species to rapidly split into a large number of taxa. The second possibility is a bit more worrying as Guyer and Slowinski's results would also be expected if the trees obtained by systematic studies were just selected at random from all possible trees. Clearly, more detailed work on this subject would seem worthwhile.

Another aspect of this type of study is the potential of phylogenetic analysis to throw light on the evenness or otherwise of speciation and extinction events across taxa. If different evolutionary lines are subject to preferential levels of either of these events then quite different tree topologies might be expected. However, before such comparisons can be made it is essential that we have some reliable way of determining the accuracy of our phylogenetic estimates.

CHAPTER THIRTEEN

MUSEUMS, HERBARIA, BIODIVERSITY, CONSERVATION AND THE FUTURE OF TAXONOMY

. . . even though museums are primary sources of information on species and even though we are in the 'information age', automation of museums' major sources of information on species, their collections and their libraries, lags a generation or more behind current technology.

(Feldman and Manning, 1992)

13.1 Museums and their roles

Few of the older generation of biologists (myself included), natural historians and taxonomists would not have been impressed by and marvelled at the profusion of preserved specimens on view at almost any natural history museum up until the 1960s. However, as anyone who has recently visited a big natural history museum will have noted, such displays are becoming increasingly rare and now are more or less limited to provincial establishments or museums in certain Eastern European and developing countries. Displays that simply show large numbers of stuffed, pickled or pinned specimens are now deemed to be outdated in the view of most modern museum exhibition managers who instead often attempt to present grander themes and overviews utilizing a good deal of high technology, although all too often a minimum number of specimens. While there can be little doubt that good photographs and videos of living organisms can greatly enhance the impact and meaning of a museum display, presentations that seem to attempt to minimize the number of genuine specimens on display place the visitor at an even greater distance from the organisms themselves. Modern 'high-tech' presentations may or may not be more educational but without doubt they hide much of the individuality of a collection's treasures. Further, these changes are not restricted to the more obvious public display areas but are also mirrored by changing

242 PRINCIPLES AND TECHNIQUES OF CONTEMPORARY TAXONOMY

approaches to the organization and utilization of the all important research collections that major museums hold safe.

Museum natural history collections serve many purposes not all of which might be immediately obvious. The main division generally comes between education and research. Perhaps with the odd exception of a few rare but educationally significant specimens such as dodo bones or a pickled coelacanth, specimens on public display are usually specially prepared for the purpose and are only exceptionally used in research. For most major museums, the vast majority of their specimens are stored safely out of the public gaze. For example, at the time of writing the Natural History Museum in London housed somewhere in the region of 67 million specimens in total of which some 25 to 30 million were insects, including approximately 279 000 primary types. These figures refer specifically to prepared specimens such as animal skins, pinned and mounted insects, or microscope slides and do not include a vast number of unprepared specimens such as bottles of unsorted plankton and **malaise trap** samples of insects. Clearly the number on display to the public at any one time is only a tiny fraction of this total.

Local museums tend to concentrate on material collected in their regions and act as important sources of information about current and past distributions of organisms. Because of this they also tend to be at the front end when it comes to providing information about local land use and conservation issues. Although there are always exceptions, provincial museums are generally not heavily involved in taxonomic revisions except when the work is also locally relevant. In contrast, major national museums usually have collections that are dominated by foreign material, and because their collections tend to be cosmopolitan, they serve as the main sources of material for taxonomic studies. Such studies may be carried out by staff of these museums or by workers in other institutions at a distance from them. In either case, it is normal for taxonomists to have to borrow material from many institutions and administering these loans is a major function of most museums and of great benefit to their collections.

13.1.1 *Management of museum collections*

By their very nature at any one time only a small percentage of the holdings of major museums will be under active study and likewise the holdings of many important natural history museums have in the past grown more due to serendipity than to organized planning. Of course this does not mean that they have failed to reach a level of great value and importance, but it does raise questions of how best to plan in advance their future acquisitions and activities.

A major worry now is the lack of growth in many major collections at a time when the world is losing species at an increasingly alarming rate. Indeed for some major institutions, it is positively embarrassing to be asked to accept large freshly collected samples — they simply do not have the cash, manpower or space to deal with them and so have to be selective about acquisitions even if this means turning away scientifically interesting material. Clearly there is a conflict of interest between the scientific need to preserve (not to mention conserve) vanishing organisms for future investigation and the immediate concern of many museums not to overburden themselves with material they could ill afford to handle.

The same conflict extends to the interpretation of the value of museum collections in different stages of curation. Most practising taxonomists make most use of material curated to family, subfamily or genus level because such material may be borrowed with ease and it is then a simple job to go through it and sort it to species, etc. Few taxonomists have the time to sift through the larger **accessions collections** (that is material preserved and mounted but only sorted into the most major categories) on the off-chance that their efforts will be rewarded with a few interesting finds. Even less do active researchers have the time to sort and prepare completely uncurated material such as bottles of pickled insects. Taxonomists have usually regarded these initial stages of curation to be the province of, or even the responsibility of, the museums themselves. Alas, times are changing and museums are increasingly obliged to ask taxonomists to help fund their curatorial loads. What this amounts to in effect is a change in the pathway of funding and in theory, assuming the same level of finance for museums remains in the system, the *status quo* might then be maintained. The problem is that many grant-giving agencies do not as yet consider it their role to help fund the museum that provides the essential resources for the research they otherwise wish to sponsor.

If any attempt is to be made to classify a large part of current global biodiversity then collections will swell like never before and this means that serious consideration must be given to the process of collections management. A noteworthy trend now gaining ground is the application of a scoring system that can allow museums to judge their current state of curation (McGinley, 1989). In this particular system, specimens are categorized into nine levels which reflect both the condition of preservation and the extent to which they are identified and therefore effectively available to other researchers. Thus level 1 designates material that is in immediate need of proper curation to ensure its future safe-keeping; level 2 defines material that is well preserved but unidentified and unsorted and therefore not available for loan; level 3 defines material that is sorted sufficiently for it to be useful for loan purposes (usually this means sorted to subfamily, tribe or genus); level 4 refers to identified material that has yet to be incorporated into the main systematic collection; levels 5 and 6 refer to identified specimens at two levels of curation; levels 7–9 refer to fully curated specimens and how much of their associated data is recorded. Assessing the holdings of a museum in terms of this or a similar system is an important step in that it enables priorities to be set for curation, expansion, etc. Contrary to what might at first be expected, a healthy active museum will not expect to have all its material in the top few levels but rather it would normally have a bimodal distribution with another peak at or around level 3 indicating that it is still adding new material and that this is not lingering long at level 4.

13.1.2 Museum funding

Museum collections are not the most expensive scientific institutions to maintain, by comparison for example with particle physics, they are quite cheap. Nevertheless, the view that all leading and hopefully a good number of developing countries should finance well-maintained collections out of government purses is rapidly dying. Prevailing attitudes of many governments demand that institutions be self-financing as far as is possible, and therefore that both public and research enterprises tailor their actions to maximize their revenues. For public displays this often means charging entrance fees and putting on items of wide public appeal that will bring in the crowds. However, the crowds are unlikely to have enough money to finance a sizable natural history museum with all its ancillary functions and in the developing world it could be argued that there is an even greater educational need that museums can help satisfy.

Museum collections cost money even if they just try to maintain the status quo by ensuring adequate curation of their existing holdings. The handling of new accessions, and of course any fundamental research that might be undertaken, costs even more. In 1992 for instance, housing and curating the insect collections alone of the Natural History Museum in London cost approximately £1.5 million. With current funding restrictions on museum collections around the world it is hardly surprising that many have been taking a more serious look at how they manage their collections, what purposes they serve, whether they are serving these efficiently, and whether any savings can be made. Ironically, these financial and managerial concerns have arisen at a time when there is a growing realization that many species are facing imminent extinction and that scientific demand on museum space will explode in the near future. Funding constraints may on the other hand lead museums to concentrate on those research areas that are capable of getting independent finance and this usually means restricting work to groups of medical, agricultural or industrial significance rather than addressing the more esoteric topic of biodiversity.

Many museums have traditionally provided expert services such as identification or advice on particular organisms such as how best to eliminate pests. For such services most institutions have levied no charge at all or only a token charge, while in fact the true cost of the man hours involved may have been considerable. Often just letting the museum keep any specimens it wanted out of the material being identified was considered a sufficient reward. In other words, many museums have under-rated the value of their own services by not charging. No private enterprise would consider such charity. Should museums charge commercial rates for their expertise? The answer to this must depend on the intentions of the body responsible for funding the museum. Usually this means public money either from the government or a mixture of government funding and income from visitors. It also may depend on the person or body making the enquiry. Perhaps an enquiry from a pesticide company that is likely to profit from the information it receives should be dealt with differently from one from a keenly interested amateur botanist or entomologist.

13.1.3 *Specimens and data*

The value of specimens is enhanced enormously if they are accompanied by ample data detailing such things as when and where they were collected, in what habitat, doing what, and by whom. Labelling specimens properly is a subject touched on by many practical guides to collecting and the precise details of what is expected and useful varies for particular groups of organisms. At the very least, the collection locality and date should be given. Information on the habitat type and method of collection can be extremely useful. For marine plants and animals, other useful information will usually include details of position on the shore or depth at which trawled or dredged. For material from mountainous or alpine areas an estimate of altitude is important.

Data labels should also always be directly associated with the specimens themselves, in a way that will not fade with time or wash out should it accidentally come into contact with a solvent. For insects on pins or worms in tubes of alcohol, labels should be written or printed in waterproof ink and either inserted inside the specimen bottle or impaled on the pin. Likewise with pressed botanical specimens, labels should be written on the actual herbarium sheets.

Whilst the importance of making sure that correct data are kept with all newly collected specimens cannot be overemphasized, it is important to realize that specimens are not necessarily worthless if they lack data and certainly should not be discarded solely for that reason. Any taxonomist wishing to examine as much material as possible of a taxon should not necessarily discount unlabelled material. More importantly, morphological and other studies increasingly require specimens to dissect, or investigate by some other destructive procedure, and for uncommon taxa it makes far more sense to carry out such investigations with spare unlabelled specimens than with labelled ones. After all, if an attempt is being made to work out the phylogenetic placement of a group using molecular or internal characters, it is to be hoped that the characters will not be influenced significantly by the precise origin of the material employed.

13.1.4 *Living culture collections*

Collections of living cultures of microorganisms are nothing new and the first culture collections were established last century not long after the general acceptance of their existence. Their importance is enormous for groups in which morphology provides little clue to identity, and now for bacterial systematics, the lodging of a type culture with at least one major culture collection is a prerequisite for the description of new taxa, with the obvious provisos for taxa that cannot at present be cultured. Cultures have equally valuable roles for fungi, algae and protozoa but as these are currently treated under the botanical and zoological codes, cultures cannot be regarded as type specimens.

Cultures of some groups of microorganisms may have importance for preserving their biological diversity. After all, a whole species cannot be collected, only at best a representative sample of the diversity that species displays. According to Hawksworth (1991) some 254 000 strains of fungi are currently in culture in collections throughout the world although allowing for synonymy these collectively probably represent only about 11 500 distinct species, fewer than 20% of those described to date. Nevertheless, the total time, cost and effort involved in maintaining these cultures is considerable. To preserve cultures of all the known species would cost about six times more and even that would only involve a fraction of the total number of fungal species that probably exist.

13.1.5 Voucher specimens

The **voucher specimen** is a source of considerable consternation among many major museums. In general, voucher specimens are the remains of material that has been used or referred to in research of many types including ecological surveys, phylogenetic studies, pest control programmes, and more. It is an unfortunate but rather common occurrence that the identities of organisms involved in published studies are called into question. To list but a couple of possibilities, the taxon worked on may turn out to be more complicated than previously realized, or future work may fail to corroborate earlier findings suggesting that the original material may have been misidentified. Voucher specimens can therefore enable the identities of organisms involved in research programmes to be checked should some future need arise. Of course, scientifically speaking, this is a laudable process which might allow future workers to resolve conflicts in their data and perhaps enable new and important results to be obtained.

To ensure their safe keeping, voucher specimens should be lodged for safe and permanent record in an established museum or other institutional collection along with full details, and their location should likewise be recorded in the original publication for cross reference. The problem with voucher specimens is the potential of many studies to produce vast numbers of them, all demanding good long-term curation. The question must be asked therefore, whether or not it really is desirable to tie up limited museum resources and valuable museum space in this particular aspect; to ask whether maintaining voucher collections is cost effective. Without doubt, voucher material is not frequently consulted in practice, although when it is, it can be very illuminating. There can be no definitive answer to this problem and views vary greatly. In a hotly debated article, Clifford *et al.* (1990) suggested that virtually all botanical voucher specimens could be pulped at no great loss to science, and that the money saved through removing the need to maintain this material could be redirected to more botanical research. These authors maintain that biology is unique in requiring that workers retain the remains of their raw materials for further study and cite the fact that few chemists would consider keeping samples of all the chemicals they have created during their work. In contrast, Diamond (1990) points out the usefulness of museums holding and maintaining series of specimens from different dates and localities. In his example, the polymerase chain reaction was employed in a museum to amplify transfer RNA genes from kangaroo rat skins that had been collected over a considerable number of years and at several localities. The results, showing spatial and temporal variation in gene frequencies, exemplify the unforeseeable use to which 'spare' museum specimens can be put.

It is also worth noting that type specimens are really only a special class of voucher specimens, but ones that are the subject of many special rules and recommendations in the various codes of nomenclature. Perhaps because taxonomists and museum administrators are so involved in type specimens, they tend to devalue other sorts of voucher specimens. Surprisingly, there are at present no rules in the ICZN or ICBN that type specimens must be deposited in major national collections, only recommendations, though proposals have been made to change this.

Perhaps in this day and age of devolved research budgets, the best solution to the voucher specimen problem would be for researchers when they are applying for funding for ecological, behavioural or other such studies that are likely to warrant the maintenance of voucher specimens, to apply at the same time for funds to cover the real costs of housing and curating their voucher material for some suitable period of time. Thus, when a museum or similar institution is asked to house voucher specimens, instead of being a burden whose associated costs detract from other things that the museum might wish to do, they would bring in an income that would help pay for appropriate curatorial staff, their training and specimen housing. However, if this situation were to develop, there would be an onus on those bodies who fund biological research to pay the costs of voucher specimen maintenance rather than considering them as disposable items of expenditure.

13.2 The future of taxonomy

Beyond the issue of museums, lies a more general question about whether taxonomy and taxonomic research will continue to be taught and funded or allowed to continue a decline that has been going on for many years (Hawkesworth and Bisby, 1988; Claridge and Ingrouille, 1992). Taxonomy has all too often been regarded as a service science. That is, it provides names for organisms so that others can refer to them, keys so that others can identify what they encounter, and classifications which facilitate data retrieval. Likewise it can produce phylogenies that are of fundamental importance for other areas of biological science in which the comparative method is central to experimental design and data interpretation.

In recent times, two factors in particular have highlighted the importance of continuing taxonomic research and producing accurate phylogenies. These are the great and increasing rate of extinction being brought about by human activities such as deforestation particularly of the species-rich tropical forests, and the realization from a number of studies that only a small percentage of living species have as yet been described (Erwin, 1982). Thus, one of the consequences of mankind's influence on the worldwide rate of organismic extinction is that vast numbers of species are going and will continue to go extinct well before any taxonomist would ever be likely to discover or describe them. In fact, given the vast number of species of organisms inhabiting the earth, with estimates frequently in the range 5 to 15 million (May, 1988, 1992), it seems likely that the great majority of living species and perhaps even genera will never be described.

Soulé (1990) has recently argued that taxonomy research should therefore be assigned sets of priority objectives amongst which might be counted medically important groups, potential economically important ones and so on. In this light, Hawksworth (1991) has pointed out that conservative estimates indicate that the number of extant species of fungi may be in the region of 1.5 million compared with only 69 000 or so described to date. Of the latter only some 7% of described species have been studied from the point of view of their secondary metabolites, yet those discovered so far include penicillin, cyclosporin and the relatively new avermectin insecticides. The potential of others to yield immensely profitable new products is obvious and as Hawksworth put it, the 'world's undescribed fungi can be viewed as a massive potential resource that awaits realization'.

One other possible priority goal suggested by Soulé might be to sample, preserve, describe, document, etc. a representative sample of the world's fauna and flora before too much has been lost. Such a survey would at least provide a valuable insight into the ranges of taxa that have evolved on our planet and probably a great deal more besides in terms of our understanding of the complexity and stratification of various communities. If this were to be accepted as a reasonable objective, it would be pertinent to consider just how representative is the sample of organisms that have already been catalogued and described. A brief moment of thought might suggest that for some poorly known tropical faunas, it is quite likely that the presently available sample will be more than a little biased towards the large pretty organisms, e.g. the big butterflies will almost certainly have been overworked compared with the smaller 'microlepidoptera', the big game over the rodents, etc. However, consider the better known faunas or floras, such as those of north west Europe. Again here we might expect our Victorian forefathers to have dealt with bigger more conspicuous plants and animals first, but surely progress has been made a long way beyond that by now. Unfortunately, according to Gaston (1991) who has recently analysed the dates of description of the British beetle fauna in relation to the insects' size, it seems that this is not the case. That is, the trend to describe bigger organisms earlier does not just include the very big and obvious taxa but continues all through the group's taxonomic history to the present day, even after more than 200 years of research.

Why has our work been so size dependent? It is easy to see, as Gaston pointed out, why workers long ago had to concentrate their attentions on larger taxa, for even their optical microscopes were not up to present day standards and they did not have access to the scanning electron microscope now so important in many fields of taxonomy. Our predecessors also had less efficient collecting apparatus and techniques than are employed today. In entomology, for example, the modification of sweep nets and the development of various flight intercept traps (e.g. **malaise traps**) has shown that it is now possible to collect prodigious numbers of specimens with comparative ease, and together with these numbers come more and more new taxa (Noyes, 1989). Perhaps the one certainty is that the taxonomist will have more than enough work to do for a very long time to come.

Another aspect of taxonomy that has a bearing on its future is how much of it is taught to new generations of students (Bramwell, 1989). A recent enlightening survey has highlighted not only the relative decline in taxonomy as a research topic but, at least in the UK, as a subject taught in higher education (Claridge and Ingrouille, 1992). At the same time advances in computing and phylogenetic theory are having dramatic effects on the actual work of systematists. The result could be that while demand for systematics might grow, the supply of adequately trained new taxonomists might not be enough to meet even present demands. If such trends are not stemmed soon, the decline in taxonomic training might even become irreversible because there would not be enough competent taxonomists to carry on training the next generation. It is to be hoped that this view is overly pessimistic.

A final point relates to the groups that taxonomists work on at present, compared with the amount of taxonomic work that particular groups require. This issue was recently examined by Gaston and May (1992) who compared the numbers of taxonomists working on particular groups with estimated numbers of described and undescribed species in each group. Not surprisingly, they found a wide range of taxonomic effort on different groups. For example, in North America approximately 32% of taxonomists work on terrestrial vertebrates (excluding fish) with an identical percentage working on insects and spiders despite the fact that in North America insects outnumber terrestrial vertebrates by a factor of at least 50. They also reported that only 2% of taxonomists worked on microorganisms although the numbers of species of microorganisms could well be staggeringly large (e.g. Hawksworth, 1990).

13.2.1 *Biodiversity and conservation priorities*

Biological diversity is so great that at present we really have no idea exactly how many species exist in the world. Without doubt many will go extinct between completion of this book and its publication, and most of these will probably never be known to science. For some groups we have a far better idea of the number of extant species and these may be used to estimate the magnitude of the current crisis. Probably the best known group of organisms due to their size, visibility and a worldwide army of enthusiasts are the birds of which some 9000 species are known to science with only a handful of new species being described each year. Of the known species, of the order of 1% are known to have gone extinct in recent times, largely as a result of human action. While this figure may sound comfortingly small, it should be noted that closer to 5% are currently under threat. In broader terms, some estimates suggest that as many as half of all species alive in 1990 will have become extinct by the year 2050 or, for example, between 25 000 and 100 000 species of plants may go extinct during the last decade of the present century (see Bramwell, 1989). Unlike the birds, however, only about 10% (give or take) of plants, fungi and invertebrates have been described.

Several arguments have been proffered to add weight to the need to conserve threatened species and habitats. Noteworthy among these are arguments concerning the potential value that the genetic make-up of many species may have for future human medical, industrial or agricultural needs. In addition of course, there is the purely esoteric benefit that knowing that they still exist in safety may give us. For example, the conservation of tropical plants ought to be an area of high priority from the point of view of the pharmaceutical industry, which in the USA alone is worth nearly $10 billion per annum with approximately 25% of prescription medicines containing active ingredients of plant origin (Farnsworth, 1988). However, only a handful of North American companies are actively screening the potential wealth of new compounds in tropical plants, and most of these are interested more from the pesticide point of view. In the same volume, Wilson (1988) gives particular emphasis to the genetic potential of the living world by considering how many separate genes are present in each organism. Thus each species of bacterium has of the order of 1000 genes while each bird or mammal may have 100 times as many. Only a very tiny fraction of these have been examined from a commercial point of view.

Irrespective of the reasons for wanting to conserve threatened habitats and species, it is certain that conservation will frequently be at odds with the economic and land needs of the local populations, and that as a result many valuable areas will be lost. It is therefore important that conservation efforts be directed most strongly to those areas that most deserve preservation. The problem is how to decide which areas are more important than others. Here systematics is playing an increasing role. For many areas we know very little about the species occurring and even when such information is currently available it has frequently been misused. First of all, it is necessary to consider what criteria should be employed in deciding conservation priorities. Is the total number of species the most important consideration, or should information about the genetic diversity of the included species be employed?

As pointed out by Stace (1980), taxonomy has at least as important a role in deciding conservation priorities as have ecological considerations, and it should be noted that an ecologist's list of priorities may not be the same as a taxonomist's. For example, on a diversity basis a taxonomist might rank highly some marginal or disturbed areas which an ecologist may prefer to ignore. However, it is doubtful that anyone would rank the conservation of the giant panda equally with that of any of the many species of rats or mice in the world so it is clear that people in general do have conservation priorities.

Recently, several methodologies have been proposed that are aimed at obtaining explicit recommendations for conservation. Given that conservation priorities will have to be set, Vane-Wright et al. (1991) devised a method of identifying a subset of extant taxa that collectively best represent the phylogenetic diversity of the group. The basis of this selection is the topology of an accepted phylogenetic tree and the method introduces the notion of close-to-baseness. The effect of applying their criteria is to choose a set of taxa that collectively represents as many of the main evolutionary lines that have evolved from the common ancestor of the group under consideration. In practice their preferred method makes use of the number of groups to which a taxon belongs, out of the total number of groups that can be obtained by partitioning the data set which in turn is related to the information content of the classification.

Vane-Wright et al.'s proposals have sparked a great deal of interest both because their methodology is a novel approach to a rapidly growing problem and because the results provide hard numbers that can be shown to legislators and conservation managers. However, theirs will not be the last word on the matter, rather they have opened up an area of great potential but one not without its own controversies. The principal source of debate is philosophical. Before we employ any methodology we should be sure that it will provide the information we actually want. Faith (1992), for example, has questioned the idea that close-to-baseness is the best criterion to employ in deciding conservation issues. Instead, Faith proposes a system based on 'phylogenetic diversity' which, rather than considering just the topology of the evolutionary tree, also makes use of the amount of genetic differentiation that the taxa display. In this scheme the group of taxa selected for the highest conservation priority should be those which collectively represent the greatest overall genetic diversity. This is illustrated in Figure 13.1 which for simplicity only indicates unique apomorphies along the evolutionary branches. Considering first the upper part of the figure, the eight taxa (A–F) collectively display 37 uniquely evolved characters. Thus, if it was only possible to save two of these eight taxa then application of Faith's procedure would suggest that conservation effort would be best concentrated on taxa B and E which together display 19 out of the 37 unique characters. If instead it was possible to conserve three taxa, then the best choices would be B, E and H which collectively represent 26 of the unique characters representing 70% of the total number in only 38% of the taxa.

Vane-Wright *et al.* and Faith's methods both have their strong points and the question of which to use boils down to what exactly the justification for conservation is perceived to be. The two procedures will often give similar results but situations can be envisaged when they will not, and this might be particularly important if a group derived from near to the base of the evolutionary tree has not changed greatly compared with more recent and rapidly evolving ones. Thus while in the upper example in Figure 13.1, the best three taxa on Faith's criterion also seem to be well spread across the tree, the same is not the case for the lower example in the same figure. In this latter case, selection of up to four taxa using Faith's method still does not provide any representatives of the least derived clade (containing taxa G and H), and the 'best choice' taxa are largely the result of their own autapomorphies.

Ultimately, however, a great deal depends on realizing that something needs to be done. After all, given that the vast majority of living organisms are unknown and are likely to remain so for the foreseeable future, many will undoubtedly slip through the conservation net unnoticed. While most of these will be small organisms, a surprisingly large number of bigger ones are far from perfectly understood and this category even includes some organisms of undoubted economic importance. For example, many of the large tropical rainforest trees so prized for their timber are still little known

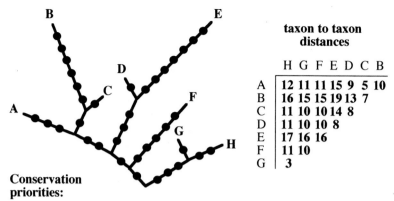

taxon to taxon
distances

	H	G	F	E	D	C	B
A	12	11	11	15	9	5	10
B	16	15	15	19	13	7	
C	11	10	10	14	8		
D	11	10	10	8			
E	17	16	16				
F	11	10					
G	3						

Conservation priorities:

if **2** taxa to be conserved then select **B** and **E** (spanning length = 19)

if **3** taxa to be conserved then select **B**, **E** and **H** (spanning length = 26)

etc.

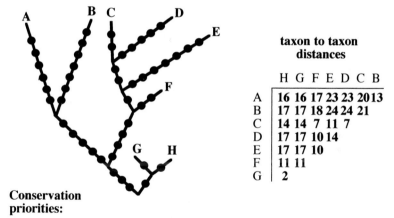

taxon to taxon
distances

	H	G	F	E	D	C	B
A	16	16	17	23	23	20	13
B	17	17	18	24	24	21	
C	14	14	7	11	7		
D	17	17	10	14			
E	17	17	10				
F	11	11					
G	2						

Conservation priorities:

if **2** taxa to be conserved then select **B** and **E** (spanning length = 24)

if **3** taxa to be conserved then select **B**, **E** and **D** (spanning length = 31)

if **4** taxa to be conserved then select **A,B**, **E** and **D** (spanning length = 37)

Figure 13.1 Two applications of the technique of Faith (1992) for assigning conservation priorities to taxa on the basis of the proportion of total variation that different subsets of taxa represent. Two different phylogenies are shown, each with the acquisition of unique apomorphous characters indicated by black spots. Counting the number of spots on the minimum spanning tree for a given set of taxa gives the measure to be maximized in deciding priorities. The heavy dependence of this technique on the number of autapomorphies is shown by the lower example in which a member of the least derived group (G and H) is only included when five out of the eight taxa are recommended for conservation.

systematically and thus if some useful species are endangered they could easily go extinct before anyone realized there was a threat. Even when there is an active desire to conserve, there is still a need for proper taxonomy.

Under a challenging headline in *Nature*, 'Profiting from Biodiversity', Nee and Harvey (1990) describe an experiment in conservation management being carried out by the government of Costa Rica under the guiding light of Dan Jansen. Costa Rica is a small and certainly not rich country with a rapidly growing population and many of the other common problems of developing countries. Nevertheless, it has boldly placed nearly 25% of its land surface in a series of national parks that collectively protect a good deal of the habitat diversity to be found in that country. Of course, with growing demands for agricultural land any government may be placed under pressure to reschedule national parks. This is where Dan Jansen's plans will hopefully provide the necessary financial arguments in favour of maintaining the parks and forests and exploiting them for their biodiversity rather than at the expense of it. The first part of the scheme involves the creation of an inventory and reference collection of the plants, animals and other organisms found in the national parks. In a tropical country such as Costa Rica with its enormous diversity this is a major task, but one of the unique features being employed there is the training and employment of teams of **parataxonomists** to build the collections that will be housed in their recently formed National Biodiversity Institute. Parataxonomists are people with no previous higher education in biology or taxonomy but who receive a basic training in collection techniques, specimen preparation and identification and are then charged with the task of assembling collections of their particular groups of organisms which can then be submitted to relevant experts for identification. Use of local labour in this way greatly reduces the costs of such projects, and evidence suggests that the parataxonomists have been fulfilling all expectations of their trainers. Another important feature of the scheme is that the main identified collections will be maintained in Costa Rica itself in purpose built institutions and not spread over institutions throughout the world. Why then, should Costa Rica be putting so much effort into organising its conservation policy in this way? Certainly the Costa Rican government has been enlightened in this respect but they have also been convinced that their natural heritage undoubtedly includes many species with marketable properties and what they want to do is to ensure that if anyone is going to make a profit from, say, a new antibiotic derived from a Costa Rican plant, then Costa Rica should get its fair share of it. What is even more important though, is that Dan Jansen and Costa Rica's collective enterprise in initiating the major conservation and biodiversity research programme means that they have justified a right to profit from any new discoveries far more than any country that simply wants to lay claim to profit based on material collected there by outsiders and studied years later.

13.2.2 *Taxonomy, zoos and captive breeding*

Taxonomic studies are beginning to have an unexpected impact in the maintenance of zoo animal populations. It has long been realized that animals which do not normally mix or interbreed in the wild do so with annoying regularity when the opportunity arises in the confines of a zoo. However, while it is obvious that lions and tigers probably ought to be separated (to prevent the occurrence of too many tions or ligers) recent cytotaxonomic studies have shown that even within the orangutans there are different chromosomal races or subspecies (Benirschke and Kumamoto, 1991). Without such information on which to base a sound captive breeding programme it is easy to imagine that both natural diversity could be inadvertently lost or even worse, crosses between unrecognized, rare sibling species might produce infertile offspring and hasten their extinction. To highlight the problem, it has only recently been recognized that the African warthog, which fortunately is in no imminent danger of extinction, actually comprises two species. If such a large animal has not been thoroughly investigated taxonomically until recently what can we expect of the myriads of smaller animals and plants?

13.2.3 *Taxonomy and legislation*

With the advent of Red Data Books and national and international conservation legislation such as the CITES agreement, taxonomy is gaining a whole new perspective, that of forming the basis for legal decisions. As a consequence of this, situations have already arisen where taxonomists have been placed in the witness box and cross-examined about all manner of taxonomic matters from details of nomenclature and the validity of subspecies and species to the justification for conservation of particular taxa. Geist (1992) provides several examples from recent cases involving large mammals where there is a surprisingly large amount of controversy concerning the status of specific and infraspecific taxa. In some cases, legislation sets down considerable penalties for collecting, hunting or trading in particular species and consequently there is considerable pressure on taxonomy to be able to provide a sound and incontrovertible framework for the legislation that will in turn help to identify and protect the taxa under threat while still permitting legal hunting, collecting, trading or land-use to continue.

Unfortunately, abundant examples indicate that legislators often do not have a thorough understanding of taxonomy, and perhaps even more so, of the difficulties that are often involved in making taxonomic decisions and identifications. Ayensu (1984) notes that US federal legislation for the protection of endangered plants and animals does not deal evenly with the two. Thus for animals the term 'species' is legally defined to include both subspecies and indeed any distinct local population, whereas no such fine divisions are allowed for plant taxa. Concerning the practicalities of enforcement Ayensu also points out that whilst laws are in place to protect many species of cacti and orchids, these are often nearly impossible to identify when they are not in flower which is precisely the condition they are most likely to be in when transported.

An interesting example of the relationship between taxonomy and protective legislation involves the tuatara, an endangered reptile from New Zealand, or more precisely a number of small islands off the coast of New Zealand. Tuataras are often thought of as 'living fossils' because they are the sole surviving representatives of a whole order of reptiles, the Rhynchocephalia, which were once the dominant group of terrestrial vertebrates. In the nineteenth century three species of these animals had been described and were recognized by herpetologists. Subsequently, one became extinct and the surviving two, *Sphenodon punctatus* and *S. guntheri*, were synonymized. Thus when the plight of the remaining populations came to be recognized and a law was passed to help protect 'the' tuatara, the authorities of the time merely followed suit and the resulting legislation dealt specifically with *S. punctatus*. However, work by Daugherty *et al.* (1990) employing morphological criteria and additionally allozyme data (see chapter 10) has clearly shown that the remaining tuatara populations do not represent a single species but rather two, as were originally recognized by Victorian herpetologists. Had the new studies not been carried out when they were, it is easy to imagine how the few populations representing the rarer *S. guntheri* could have been allowed to slide into extinction. This would have been a bad blow to herpetology and disastrous for one species of tuatara. As pertinently put on the front cover of *Nature* which reported the rediscovery of the second tuatara, 'Bad taxonomy kills'.

Even if courts do concur with taxonomists about the identity of taxa covered by protective legislation, the species may themselves conspire to confuse matters in that, especially among certain plants, birds and mammals where current legislation is concentrated, species and more often subspecies may hybridize and backcross leading to genetic introgression and loss of purity. In terms of the law then, the question ends up being asked whether a specimen purportedly belonging to one taxon can in fact be legally regarded as a member of that taxon (and hence covered by the law) if it possesses some genes belonging to a related subspecies or species since it would then no longer be pure. So far such matters depend on the decisions of courts in each particular instance and consequently leave many potential loopholes unclosed and therefore liable to exploitation.

GLOSSARY

The following list includes terms that will be encountered frequently in both taxonomic works and those dealing with the methods by which phylogenetically informative data are obtained such as immunology, histology and biochemistry. To increase its usefulness many entries give references to relevant publications included in the Bibliography. Nomenclature, in particular, has accumulated many specialized terms of its own and no attempt has been made to include a comprehensive list of these here although many of the more frequently encountered are presented. Jeffrey (1989) provides a more detailed coverage.

Aberration   A typically uncommon, variant on the typical form of a species; aberrations are treated as infrasubspecific and therefore have no status in the **Codes** (q.v.).

Accessions collection   Material held in museum collections that has not yet been accurately sorted and incorporated into the main collection.

Acentric   A chromosome without a centromere.

Additive coding   See **ordered character**.

Agamospecies   Any plant which obligately (or habitually) reproduces by means of **agamospermy** (q.v.) and which, as a result, forms genetically isolated 'microspecies' (Mayr, 1992).

Agamospermy   Any of many mechanisms by which a plant sets seed without fusion of gametes.

Agglommerative clustering   Any clustering procedure that produces groups of entities (i.e. OTUs) by progressively linking similar ones in the same group, cf. **divisive clustering** (see section 4.3).

Allele   Any of one or more different variants of a gene at a given gene locus.

Allopatry   Pertaining to taxa or populations that occupy geographically separated areas.

Allopolyploidy   **Polyploidy** (q.v.) that results from the combination of a chromosome set from each of two different individuals or more usually taxa (cf. **autopolyploidy**).

Allotype   A member of a type series of a species representing the opposite sex to that of the holotype; allotypes are selected by the original author who believes them to belong to the same species (the term is not recognized by the Codes).

Allozyme Any of one or more variants of an enzyme (usually motility variants identified by electrophoresis) coded by different alleles at the same gene locus (Avise, 1974; Richardson *et al.*, 1986).

α-taxonomy The process of naming and defining species (Mayr, 1969).

Amphidiploid An organism of polyploid hybrid origin which now behaves as a diploid.

Anagenesis The (gradual) accumulation of genetic change over evolutionary time (Horner *et al.*, 1992; see Figure 12.1).

Analogy Refers to superficial similarity between two characters or character states that does not reflect common evolutionary origin; cf. **homology**.

Aneuploidy State when chromosome number varies from normal ($2n$) by some amount other than an integer multiple of a haploid chromosome set (cf. **euploidy**).

Annealing The process by which two homologous or near homologous single-stranded nucleic acid strands align and bind to form a double-stranded nucleic acid.

Anticodon Nucleotide triplet in tRNA which pairs with a **codon** in mRNA.

Antigen A molecule capable of causing the production of specific antibodies by the vertebrate immune system.

Antigenic determinant (or **antigenic site**) The site on an antigen molecule that is recognized by a particular antibody; typically such sites comprise between 5 and 10 amino acids.

Apomixis Widely used to mean **agamospermy** (q.v.).

Apomorphy An advanced or derived character state (cf. **plesiomorphy**).

A posteriori **weighting** A method for differentially weighting characters based on their fit to an initial estimation of tree topology obtained without weighting (Farris, 1969; Legendre, 1975; Carpenter, 1988).

A priori **weighting** Differential character weighting not relying on an initial phylogenetic estimation, either using an arbitrary system of weights or deriving weights from compatibilities (Neff, 1986; Sharkey, 1989).

Autopolyploidy Polyploidy (q.v.) resulting from the doubling of the chromosome number within an individual (or its gamete) (cf. **allopolyploidy**).

Attribute Another name for **character state** (q.v.).

Autapomorphy A derived character unique to a single taxon, cf. **synapomorphy**, **symplesiomorphy**.

Autoradiography Process for localizing sites of radioactivity in macro- to microscopic specimens based on closely apposing a radiation sensitive (photographic) plate or emulsion to the subject.

Availability In relation to a taxonomic work, whether its contents can be considered as valid taxonomic acts under the requirements of the appropriate Code.

Available names A scientific name that can be considered as a possible valid name for a taxon, i.e. it is not excluded by any factors specified by the appropriate Code.

Axenic   Pure, uncontaminated, in reference to living cultures.

β-taxonomy   The process of arranging taxa into higher categories that reflect evolutionary history (Mayr, 1969).

Bacteriophage   See **phage**.

Binomen   The combination of a generic name and a trivial (species) name which comprises the full scientific (Latin) name of an organism.

Biodiversity   A widely used term (contraction of biological diversity) that can be taken to refer to the diversity (number) of taxa and/or their combined genetic variation, or more generally to mean almost all aspects of biology.

Biological species   A definition of 'species' based on the principle that gene flow occurs readily and frequently between members of a species and not or hardly at all between different species (cf. **phylogenetic species**). Biological species definitions run into problems with non-sexually reproducing organisms (Mayr, 1942, 1992; Mayr and Short, 1970).

Biovar or **biotype**   In bacterial taxonomy an infrasubspecific rank of no official standing denoting a strain distinguishable on the basis of biochemical or physiological properties.

Blot   A sheet of nitrocellulose, nylon or similar material on to which an electrophoretically separated pattern of proteins, DNA or RNA have been transferred either by passive diffusion or electrophoresis perpendicular to the original separation; blots enable further procedures (e.g. immuno-labelling, probe-binding, autoradiography) to be carried out on the separated macromolecules.

Bootstrapping   A data randomization procedure used to assess the confidence that can be placed in the results of an analysis of a particular data set (Felsenstein, 1985b).

Branch-and-bound   An algorithm guaranteed to find minimum length trees but often requiring considerably fewer calculations than **exhaustive search** (q.v.) (see section 3.2.1).

Caminacules   A set of hypothetical organisms created by Camin during the 1960s using an evolutionary model to derive each caminacule from its ancestor. The resulting 'extant' caminacules were presented to colleagues to see if they could reconstruct their phylogeny.

Camin–Sokal character   Another name for irreversible character (first used by Camin and Sokal, 1965) (see section 2.2.5).

Canonical variate analysis   A powerful discriminant procedure much used in identification (section 4.4.3).

Caryotype   See **karyotype**.

C-banding   A chromosome banding technique using **giemsa** (q.v.), so-called because originally it was found to stain centromeric heterochromatin. It predominantly reveals satellite DNA (Macgregor and Varley, 1983).

Centric fusion   Process in which two separate chromosomes become fused with the loss of one of the centromeres to form a single larger chromosome (also known as Robertsonian fusion).

Centroid   The point in a multidimensional space with each axis representing a character variable, which is defined by the mean value of each character for a group of taxa or OTUs.

Centroid clustering   A hierarchical clustering method using distances between group centroids as the distance measure (section 4.3.5).

Centromere   A defined part of a chromosome where the two sister chromatids are attached during cell division.

Centrotype   The most typical strain of a bacterial species in culture. This has no special standing in bacterial nomenclature.

Character   Any physical structure (macroscopic, microscopic or molecular) or behavioural system that can have more than one form (**character state** q.v.), the variation in which potentially provides phylogenetic information.

Character state   Any of the possible, distinct conditions that a character can display. Sometimes also loosely termed character.

Chemotaxonomy   The process of obtaining and applying taxonomic information about chemicals produced by and chemical pathways of organisms, although generally taken to exclude sequence data from proteins and nucleic acids and other properties of macromolecules.

Chromatin   The combination of DNA and associated proteins (histones) that make up the bulk of nuclear and chromosomal material.

Clade   Any supposedly monophyletic group of taxa in a phylogenetic hypothesis (Huxley, 1958; Dupuis, 1984).

Cladist   An advocate or practitioner of **cladistics** (q.v.) and its principles (Mayr, 1965; Dupuis, 1984).

Cladistics   In general, the process of defining evolutionary relationships between taxa using evidence from extant taxa. Originally formulized by Hennig, now comprising a number of variously related methodologies (Fitch, 1984).

Cladogenesis   The evolutionary process giving rise to a new **clade** (q.v.).

Cladogram   A dendrogram (tree diagram) specifically depicting a phylogenetic hypothesis and therefore based on synapomorphies. A cladogram generally only indicates the branching pattern of the evolutionary history, cf. **phylogram** (Mayr, 1965; Wiley, 1981).

Class   A formal classificatory group between phylum and order (also stem of subclass and superclass).

Classification   The aspect of taxonomy concerned with arranging taxa into a classification.

Cline   A continuous spatial gradient in the frequency of a particular **allele** (q.v.).

Clique   A set of compatible characters. See **compatibility analysis** (see 3.1.2).

Clone   (1) A group of genetically identical organisms resulting from non-sexual cell division processes; (2) more loosely, a number of copies of a fragment of DNA produced by **cloning** in a microorganism (see chapter 11).

Cloning   A procedure for amplifying the number of copies of DNA sequences by incorporating them within a bacterium where they are replicated in large numbers (see section 11.3.1).

Cluster analysis   Any procedure that progressively links entities (i.e. OTUs) into clusters on the basis of their attributes (e.g. overall similarity). Usually used only to refer to techniques employed in phenetic rather than cladistic analysis.

Codes   Either of the **ICBN, ICNB, ICNCP** or **ICZN** (q.v.) which formally regulate the application of names to their respective groups of organisms through their rules and additionally provide recommendations and principles. At present the ICTF only has a recommendatory role.

Coding   The assigning of usually numerical values to different character states such that they can be subjected to mathematical manipulations.

Codon   Nucleotide triplet in mRNA coding for an amino acid, cf. **anticodon**.

Cohort   A classificatory group between family and genus usually restricted to botanical classifications (but see Wiley, 1981).

Commonality principle or the **common equals primitive method**   The idea that plesiomorphous character states will be commoner on average than apomorphous ones (Estabrook, 1977; Stuessy and Crisci, 1984; Frohlich, 1987).

Compatibility analysis   Any of several procedures for eliminating characters from a phylogenetic analysis on the basis of their not being consistent with other characters (see Le Quesne, 1964; Estabrook *et al.*, 1977; Meacham, 1984; Meacham and Estabrook, 1985; also sections 3.1.2, 3.3.2).

Complement   A complex set of immunologically active plasma proteins in vertebrate blood that interact with antigen–antibody complexes. Used in the **microcomplement fixation** (q.v.) assay because of their ability to lyse red blood cells that have been sensitized by coating them with appropriate antibodies.

Congeneric   Of two or more species believed to belong to the same genus; hence congeners.

Congruence   Degree of similarity between two phylogenetic or classificatory systems, two identical classifications being said to show perfect (100%) congruence. Often applied to comparisons of trees derived from different data sets (Rohlf, 1963; Mickevich, 1978; Colless, 1980; Crisci, 1984; Sackin, 1985; Rohrer, 1988; see section 3.3.3).

Consensus tree   A tree displaying as resolved only those features that are the same in all (or in some cases, the majority) of a set of trees suggesting different phylogenies for the same set of taxa (Adams, 1979; Smith and Phipps, 1984; Bremer, 1990; see section 3.2.8).

Consensus index   Any of a group of numeric descriptors of the information content of a consensus tree (Rohlf, 1982; see section 3.2.8).

Consistency index   A number describing the fit of a single character or of the whole set of characters (more fully the ensemble consistency index) on a tree (see section 3.2.5).

Consortium A name applied to an association of two or more organisms, especially in bacteriology.

Cosmid A **plasmid** (q.v.) into which the *cos* gene of the lambda **phage** (q.v.) has been inserted. Used for cloning large lengths of DNA (*c.* 25–40 kilobases) in *E. coli.*

Cotype A paratype or syntype. An obsolute usage no longer recognized in zoological nomenclature.

cpDNA Chloroplast DNA.

Cryptic character A character those states are difficult to observe perhaps because of small size or hidden location.

Cryptic species Members of pairs or groups of biological species that are difficult to define or distinguish from the other members on the basis of external morphological characters.

Cultivar or **cv.** In horticulture and agriculture equivalent to a plant (or fungus) variety (Stace, 1980; Jeffrey, 1989).

cv. Cultivar (q.v.).

Deme Usually a small, local subunit of a species' total population, individuals of which form a more or less isolated gene pool.

Designation (1) The act by a revising taxonomist whereby a single specimen from a series of paratypes, syntypes or cotypes is selected as the lectotype; (2) the subsequent selection as a type species of a genus of one of a number described as congeners following an original generic description in which the original author did not specify a type species. (The relevant Codes list factors to be taken into consideration when a species or specimen needs to be designated as a type.)

Determinant A region on the surface of a protein molecule (or other antigenic substance) that is recognized by a particular antibody.

Dichotomy A branch point (node) tree or a decision point in a key where two new branches arise from the stem.

Dideoxynucleotide A modified nucleotide which when incorporated into a nucleic acid sequence prevents further chain extension with DNA polymerase. Used in DNA sequencing.

Digestion Enzymatic treatment of a macromolecule (e.g. DNA, RNA, protein) that results in a number of smaller cleavage products. Increasing sequence specificity of the enzyme will yield fewer larger fragments.

Diplotene A stage during the first meiotic cell division in which chiasmata formation and crossover occur.

Directed character The same as **polarized character** (q.v.); a character whose ancestral state is specified in a phylogenetic analysis. Opposite is **undirected character**.

Discriminant analysis A weighted cluster analysis method aimed at maximizing the likelihood of correctly identifying an unknown (specimen) belonging to one of two or more difficult to distinguish taxa.

Dissimilarity Any distance measure which gives 0 for identity and a non-zero value related to the amount of difference between two taxa (= distance).

Distance matrix A taxon-by-taxon matrix of distance or dissimilarity measures (see section 4.1).

Distance Wagner method A progressive tree construction algorithm using taxon–taxon distances and employing any of several optimality criteria to select which taxon is added to the tree at each stage of construction (Swofford, 1981; see chapter 3).

Divergence coding A method for converting continuously variable characters into discrete codes for phylogenetic analysis (Thorpe, 1984; see section 2.2.1).

Division An informal classificatory group between class and order (or **cohort** q.v.) principally in zoological works (Wiley, 1981). Also **divisio**, a formal classificatory group between kingdom and class in botanical nomenclature (also stem of subdivision) equivalent to **phylum** (q.v.) in zoological classification.

DNA fingerprinting Any of several related techniques based on restriction fragment analysis (see **restriction fragment length polymorphism**) that reveal polymorphisms at dispersed loci of tandemly repeated DNA (**minisatellite DNA** q.v.) (Jeffries *et al.*, 1985, 1990, 1991). Used principally for determining relatedness between closely related individuals.

DNA hybridization Either the general process by which single-stranded DNA will associate (pair) with other homologous strands or more specifically an analytical procedure in which the relative affinities with which single-stranded DNA from pairs of organisms will pair is determined. The affinity measures in the latter case are often used to estimate genetic divergence between taxa (chapter 11).

Dollo characters Characters that are assumed for the sake of cladistic analysis only to evolve in one direction. Named after Dollo's law, i.e. that in evolution characters once lost are never regained, hence Dollo parsimony (see Farris, 1977; Felsenstein, 1981).

Domain An informal term recently proposed, with some merit, as a hierarchic level above kingdom to emphasize the fundamental difference between pro- and eukaryotes (Mayr, 1990).

Donor The organism in a DNA cloning procedure whose DNA is to be amplified.

Driver In **DNA hybridization** (q.v.), sheared pieces of unlabelled DNA to which the labelled tracer DNA is allowed to hybridize (cf. **tracer**) (see section 11.6).

Ecotype A distinct form of an organism that develops under a given set of environmental conditions.

Electromorph A variant, usually of a protein, characterized by its motility properties during **electrophoresis** (q.v.).

Electrophoresis Any of several related procedures for separating molecules in a supporting medium on the basis of their electric charge in combination with various other factors (see also **blot**, **IEF**, **PAGE**, **SDS**).

Endemic   Of a taxon, usually a species, restricted to an area in which it originated.

Epithet   In botanical and bacteriological nomenclature, that part of a scientific name following the generic (or subgeneric) name. More or less equivalent to species name but also includes subspecies names.

Euchromatin   Chromosomal material which appears weakly staining (i.e. less condensed) with DNA specific stains and includes a high proportion of coding regions.

Euclidean distance   As applied to taxonomy, the square root of the sum of squares of the differences in numerically coded characters between pairs of taxa (Sokal and Sneath, 1963; Rohlf, 1974).

Eukaryote   An organism (or cell) in which the genetic material is complexed with histones and enclosed within a membrane-bound nucleus (specifically members of the kingdoms Protista, Fungi, Plantae and Animalia).

Euploidy   State when chromosome number is changed from the normal haploid number $(2n)$ by an integer increment of n (cf. **aneuploidy**).

EU (evolutionary unit)   A term equivalent to **OTU** (q.v.) more acceptable to cladists though not widely used.

Exhaustive search   The process of finding most parsimonious trees by considering all possible trees (see section 3.2.1).

Facies   The general form and appearance of an organism or group of organisms.

False character   A character whose distribution of states among taxa suggests a phylogeny different from the true ancestry of a group and therefore one which must have shown convergence, parallelism or reversal (cf. **true character**) (see section 3.1.2).

Family   A formal classificatory group between order and tribe (also stem of subfamily and superfamily).

Family group name   Any recognized taxonomic category above genus up to and including superfamily.

Feulgen staining   Cytological staining procedure, effectively specific for DNA, based on the reaction of aldehyde groups created by partial hydrolysis of deoxyribose in DNA with decolorized fuchsin (Schiff's reagent). Can be applied quantitatively.

Fitch character   Another name for **unordered** (q.v.) or non-additive characters.

Fitch–Margoliash method   A method for constructing trees from distance data which employs branch swapping procedures after an initial tree has been constructed (Fitch and Margoliash, 1967).

Fixation   The act of specifying the type of a supraspecific taxon either in the original description (original designation) or subsequently (cf. **designation**).

Form   Lowest supplementary taxonomic rank in botanical nomenclature. Not accepted for taxonomic purposes by the International Code of Zoological Nomenclature.

Gap coding   Methodologies that convert potentially continuously variable characters into a number of discrete character states by identifying gaps of defined magnitude in the variation (Thorpe, 1984).

G-banding   A chromosomal banding technique using **giemsa** (q.v.); possibly related to degree of disulphide bridging in associated non-histone proteins and occurring primarily in AT rich chromosome regions (Seabright, 1971; Macgregor and Varley, 1983; Hillis and Moritz, 1990).

Generalized gap coding   A method for discetely coding continuously data (similar to **divergence coding** (q.v.) but different from **gap coding** (q.v.)) (Archie, 1985; Farris, 1990).

Genetic distance and **genetic similarity**   Any of several measures of the degree of divergence (or lack of it) between populations calculated using evidence for allele frequencies (Nei, 1972, 1978; Rogers, 1972; Thorpe, 1979).

Genetic drift   Random mutations and changes in gene frequency within a population that over time causes its 'average' genetic composition to depart further and further from its starting conditions.

Genome painting   A technique involving **in situ** hybridization with fluorescently labelled DNA used to identify the origins of chromosome segments in hybrids (see section 7.6).

Genospecies   The concept of a group of bacterial taxa (species) that are capable of exchanging genetic material.

Genotype   (1) The type species of a genus; an obsolete usage which should now be replaced by **type species** (q.v.); (2) the genetic composition of an organism; often used in connection with one or a few specified gene loci.

Genus   A formal taxonomic grouping that comprises one or more species that are (usually) believed to be closely related.

Genus-group name   Generic and subgeneric name.

Giemsa stain   A widely used chromosome staining mixture (including eosin Y and methylene blue) used in C-banding.

Grade   Any collection of organisms displaying a similar level of complexity (either overall or with respect to a particular system) cf. **clade**. The level of complexity itself (Huxley, 1958; Simpson, 1961).

Gradient electrophoresis   Electrophoretic technique for separating macromolecules according to their size by means of a gel whose concentration (and hence pore size) varies continuously from one end to the other (see section 10.2).

Gradualism   A concept of Darwinian evolution whereby species are believed to evolve through the gradual accumulation of small changes (cf. **punctuated equilibrium**).

Gram's stain   A staining process that enables bacteria (prokaryotes) to be assigned to one of two major groups depending on their cell wall chemistry, hence 'Gram positive' or 'Gram negative'.

Group average method   Another name for **UPGMA** (q.v.).

Hapantotype   A special concept of type used in connection with the Protozoa comprising several apparently related individuals at different stages in their developmental cycle mounted on one or more slides, etc.

Haploid In eukaryotic organisms in which reproduction depends upon the combining and subsequent halving of chromosome number, the stage displaying the smaller (halved) chromosome number.

Hardy–Weinberg equilibrium The property that allele frequencies for a given gene locus in a large population of a sexually reproducing diploid organism with **panmixis** and in the absence of differential selection will remain constant from one generation to another; for a two allele system, if the two alleles, P and Q, are present in the population in the frequencies p and q then the expected frequencies of the two homozygotes, PP and QQ, and of the heterozygote, PQ, will be given by p^2, q^2 and $2pq$, respectively (see section 10.3).

Hennigian Relating to the cladistic processes of defining taxa (clades) by synapomorphies as proposed by Hennig (1950 *et seq.*).

Heterochromatin Chromosomal material which appears darker (i.e. is more condensed) with appropriate staining, is largely untranscribed and whose DNA generally replicates later than the remainder, cf. *euchromatin*; comprises several types including concentrations of satellite DNA.

Heterochrony Modification in the timing of aspects of an organisms development that leads to changes in phenotype.

Heteroduplex Double-stranded DNA in which each of the two strands was isolated from a different taxon.

Heterologous Relating to the cross-reaction of one component derived from one species with a complementary component derived from a different species. For example, in DNA hybridization between taxa or interaction of antigen from one taxon with antibody raised against antigens from another taxon.

Heterozygosity The possession in a diploid (or higher ploid) organism of two (or more) forms of a gene at a given locus.

Heuristic (1) Of a general method that involves the progressive improvement of estimates by trial and error search rather than by following a set method; (2) of algorithms for finding shortest trees that are not guaranteed to find the most parsimonious one (section 3.2.1).

Histone Any of several basic proteins associated with the chromosomal DNA of eukaryotes. In general, amino acid sequences of histones have been highly conserved and they have therefore been useful in interpreting higher level relationships.

Holocentric Of a chromosome all parts of which appear to be pulled during cell division and therefore behave 'normally' even after fragmentation (Jackson, 1971).

Holophyletic Term employed by Ashlock (1971) equivalent to **monophyletic** (q.v.) *sensu* Hennig (Platnick, 1977; Wiley, 1981).

Holotype The specimen on which the concept of a species or subspecies is based. In most instances the holotype must be a single specimen identified as such by the original author. In the case of certain protozoa the concept may be extended to cover a number of individuals (believed to be conspecific) on a microscope slide although in these cases they should more properly be referred to as a **type slide** (q.v.) (see also **hapantotype**) (Faegri, 1983).

Homeotype   Same as **metatype** (1) (q.v.).

Homology   (1) Similarities between structures or other characters in two or more taxa that are the result of their inheritance from a common ancestor (cf. **analogy**) (Remane, 1956; Rieger and Tyler, 1979; Patterson, 1982; Humphries and Funk, 1984; Wagner, 1989); (2) also (mis)used in molecular biology to express the similarity between macromolecular sequences (Davison, 1985; Moritz and Hillis, 1990).

Homonym   Either of two or more identical scientific names that could cause a conflict of interpretation in taxonomy. In practice, homonymy applies either to species names that are at any time placed within a single genus, or genus group names within any one of the domains of the international regulatory bodies dealing with nomenclature.

Homoplasy   Generally a term used to express the sum of additional numbers of character state changes that a particular phylogeny implies above the minimum number of changes that could theoretically have taken place, given the total number of character states. Thus it includes excess changes resulting from parallel or convergent evolution and from character state reversals (Archie, 1989b; Sanderson and Donoghue, 1989; Wake, 1991).

HTU (hypothetical taxonomic unit)   Hypothetical ancestral taxa within a cladogram.

Hydroxyapatite   A mineral used in DNA studies because of its ability to bind selectively to double-stranded DNA.

Ichnofossil   A fossil of a working or construction of an organism rather than of the organism itself, for example animal tracks, burrows, nests, etc.

Ichnotaxon   A taxon based upon an **ichnofossil** (q.v.).

ICBN   International Code of Botanical Nomenclature.

ICNB   International Code of Nomenclature of Bacteria.

ICNCP   International Code of Nomenclature of Cultivated Plants.

ICTF   International Committee for the Taxonomy of Fungi.

ICZN   International Code of Zoological Nomenclature.

Idiogram   As applied to chromosomes a stylized chart of the karyotype showing relative sizes, arm-lengths, etc. (section 7.2).

IEF   See **isoelectric focusing**.

Immunoelectrophoresis   A combined electrophoretic and immunological technique in which a mixture of antigens are separated in a gel by electrophoresis and then reacted to antibody diffusing into the gel so as to form a series of separate precipitin arcs where each antigen meets its specific antibody (see chapter 9).

Immunoglobulins   Immunologically important serum proteins in vertebrates that recognize and bind to antigen molecules, e.g. IgG and IgM.

Indicator strain In bacteriology a standard strain used by various workers in lieu of a type or neotype strain (Buchanan and Gibbons, 1974).

Informative character Any character whose distribution of states among the taxa under consideration could potentially provide information about the phylogenetic relationships of these taxa. In practice this excludes invariant characters and characters displayed by only a single taxon within the group.

Ingroup The apparently monophyletic group of taxa the relationships of which are under investigation (cf. **outgroup**).

in situ **hybridization** Autoradiographic procedure for localizing specific DNA sequences in whole chromosome preparations (Macgregor and Varley, 1983). Usually only applicable to localizing genes represented in the genome by large numbers of copies (section 7.6). See also **genome painting**.

Introgression Spread of genetic information from one gene pool or species to another through hybridization and subsequent backcrossing. Extremely important in the evolution of plants but probably considerably rarer in animals.

Intron A sequence of DNA that does not code for an amino acid sequence but which is included within the DNA sequence of a structural gene. Introns are transcribed into mRNA together with the structural gene regions but are spliced out because of recognition sequences at either end before the mRNA reaches the cytoplasm and is transcribed. Presence or absence of introns can be used as a taxonomic character (see e.g. Dixon and Pohajdak, 1992).

Invariants A rationale pioneered by Lake (1987) for interpreting DNA sequence data (see section 3.1.3).

Inversion A chromosomal rearrangement in which a section of a chromosome has been excised and reinserted the other way around (see chapter 7).

Isoelectric focusing (IEF) An electrophoretic separation technique usually used to distinguish protein variants, based on their migration within a pH gradient to their **isoelectric points** where they accumulate (focus) (see chapter 10; McLellan and Inouye, 1986; Kasmer, 1991).

Isoelectric point The pH at which a zwitterion such as a protein has no net charge.

Isotype In botanical nomenclature, a duplicate of a holotype which was collected at the same time and place as the holotype (Faegri, 1983).

Isozyme Either of more than one form of an enzyme coded for by a gene at a different locus than another form of the same enzyme. May arise, for example, through gene duplication, cf. **allozyme**.

Karyotype The chromosomal complement of a cell or organism.

Key A series or set of questions requiring decisions about taxa (or other items) that lead one to an eventual identification (Osbourne, 1963; Newell, 1970; Pankhurst, 1970, 1978; Tilling, 1984, 1987; Duncan and Meacham, 1986).

Kingdom In most classificatory systems, the highest taxonomic category, normally comprising the Animalia, Plantae, Fungi, Protista and Monera

(Prokaryota: bacteria and blue-green algae). Some modern phylogenetic classifications subdivide the Monera into three major groups.

Kleptotype   An unofficial term applied especially in botany to part of a type specimen removed without consent to another collection by an unscrupulous worker.

Lambda phage   The most widely used vector for gene cloning work which infects *Escherichia coli*. Many strains are available which are suitable for cloning different lengths of DNA (see also **phage** and **cosmid**).

Lectoparatype   Equivalent to **paralectotype** (q.v.).

Lectotype   An individual specimen selected from the **type series** (q.v.) of a previously described species so as to fix the identity of that species (Faegri, 1983).

Legitimate name   In bacterial and botanical nomenclature a name that is validly published and must be taken into account when deciding matters of **priority**, **synonymy** and **homonymy** (q.v.).

Locus   A position within the genome occupied by a gene.

Lumper   An informal description of a taxonomist who tends to include groups with only slight differences within the same taxon (cf. **splitter**).

Lundberg rooting   A method for rooting a tree in which an outgroup taxon is added to a previously computed most parsimonious tree.

Malaise trap   A tent-like intercept trap for flying insects now widely used in sampling insect faunas due to its efficiency (see Noyes, 1989).

Manhattan distance   As applied to taxonomy the sum of differences in numerically coded characters between pairs of taxa (Sneath and Sokal, 1973; Rohlf, 1974).

Maxam–Gilbert method   A method of DNA sequencing based on cleaving DNA molecules at selected nucleotides and determining the lengths of the fragments produced.

Maximum likelihood   A statistical approach to constructing cladograms that seeks to find those trees that are most likely to explain the observed data. Precise definition of an evolutionary model of character state change is necessary and this generally limits its use to the analysis of molecular sequence data (Felsenstein, 1981; Swofford and Olsen, 1990; Navidi *et al.*, 1991; see section 3.1.3).

MC′F or **MCF**   Microcomplement fixation (q.v.).

Meristic character   A character whose states are determined by the number of times a component structure is repeated, for example the number of segments in an insect antenna.

Metaspecies, metataxon   A species or other taxon, respectively, which while only defined by **symplesiomorphies** is not known to be paraphyletic (Donoghue, 1985; section 2.3.1).

Metatype   (1) A specimen compared with a holotype and believed by a competent taxonomist to belong to the same taxon; (2) especially in botany, it is implied that the comparison is made by the person who originally described the holotype (cf. **homeotype**) (Sivarajan, 1991). NB In neither sense is the metatype concept recognized by the **Codes**.

Microcomplement fixation A technique for comparing the affinities of homologous and heterologous antigens for a particular antibody which makes use of the selective binding of complement to tightly bound antigen–antibody complexes (complement fixation). Unfixed complement is finally assayed by the lysis of sensitized red blood cells (virtually always sheep red blood cells) thus indirectly assaying the amount of bound complement (Champion *et al.*, 1974; Maxson and Maxson, 1986, 1990; chapter 9).

Minisatellite DNA Small, tandemly-repeated sequences spread throughout animal genomes, each repeat unit usually comprising a conserved core sequence and some more variable base positions. Important in **DNA fingerprinting** (q.v.).

Molecular clock The idea that the rate of evolutionary change in some molecule (e.g. the sequence of a particular gene or protein) is well correlated with time across a wide range of organisms and therefore, one calibrated, can be used to date evolutionary splits (Thorpe, 1982).

Monoclonal antibody A pure antibody (specific for a single antigenic determinant) produced from a specific clonal line of hybridoma cells (cf. **polyclonal antibody**).

Monophyletic Two definitions are widely employed. Used here in the same way as Hennig (1966) to mean a taxon or group of taxa all members of which have a common ancestor and which includes all the descendents of that ancestor. This corresponds to 'holophyletic' of Ashlock who also considered **paraphyletic** taxa (q.v.) to be monophyletic (Hennig, 1966; Ashlock, 1971; Platnick, 1976; Hull, 1979; see section 2.4.1).

Monothetic Pertaining to dichotomous keys in which each couplet contains a single pair of non-overlapping character-state options (cf. **polythetic**) (chapter 5).

Monotypic Of a taxon which contains only one member of a subordinate taxon; for example, a genus containing only one species or an order containing only one family (Platnick, 1976).

Monotypy A principal reason for designating the type of a taxon when one was not specified in the original description. The single originally included taxon of a monotypic taxon must be designated as the type.

Morphocharacter Short for morphological character. Generally refers to all forms of physical features from gross morphology to ultrastructure although typically excludes chromosomal features and cell biochemistry.

Morphovar or **morphotype** In bacterial taxonomy, an infrasubspecific rank of no official standing denoting a strain distinguishable on morphological grounds.

Name-bearing type A new term introduced in the 1985 **ICZN** to indicate either a type genus, type species or any accepted type specimens which provide an objective standard for the application of a scientific name.

Neotype A specimen selected to represent and fix a previously described species whose original type material is irretrievably lost or destroyed; neotypes should be carefully chosen and preferably originate from the same locality as the original holotype. The relevant codes lay down strict instructions for the selection of neotypes.

Network Usually used in the same sense as **unrooted trees** although in graph theory it may be given a broader meaning.

Nomenclature The process of assigning correct scientific names to organisms in agreement with the relevant **Codes** (cf. **classification, systematics** and **taxonomy**).

Nomen conservandum **(conserved name)** A name that would be invalid in accordance with the appropriate Code but which is made valid through the use of the appropriate Commission's **plenary powers** (q.v.).

Nomen dubium **(uncertain name)** A scientific name that cannot be associated with certainty with any known taxon because the original description is inadequate and the type specimen is lost or in too poor a condition.

Nomen hybridum **(hybrid name)** A scientific name derived from two or more separate languages (e.g. Latin and Greek). Not recommended by the **Codes** but not prohibited.

Nomen novum **(new name)** A name proposed to replace an existing name that is invalid due for example to homonymy.

Nomen nudum **(naked name)** A published scientific name (Latin binomen) used to refer to an organism which does not exist or has not been described in accordance with the relevant Code.

Non-additive character **Unordered character** (q.v.).

Normalization A mathematical procedure which converts a set of continuously variable character values to a set with a mean of zero and a standard deviation of 1.0.

Nothomorph A distinct variant of an interspecific hybrid usually resulting from separate hybridization events.

Nothotaxon A taxon resulting from hybridization of two or more taxa of the same rank. Term used principally in botanical nomenclature.

Nucleotides Nitrogen-containing organic molecules that encode genetic information in DNA and transcribe it in RNA, viz. the purines adenine (A), and guanine (G), and the pyrimidines cytosine (C) and thymine (T) in DNA, or cytosine and uracil (U) in RNA. In double-stranded nucleic acids, purines pair with pyrimidines in the combinations C–G and A–T (or A–U).

Numerical taxonomy Broadly any numerical approach to taxonomy (including cladistics) but now widely employed to refer to **phenetic** (q.v.) methods that seek to cluster and classify purely in accordance with similarity rather than phylogenetically as is the rational of cladistics (Sokal and Sneath, 1963; Sneath and Sokal, 1973; Clifford and Stephenson, 1975).

Objective synonym Synonymy where two or more named taxa (genus or species) have the same type (species or type specimens, respectively) and therefore about which there can be no difference of opinion (cf. **subjective synonym**).

Operational taxonomic unit or **OTU** Effectively any entity treated separately for the purpose of classificatory analysis but now used almost entirely in the context of numerical taxonomy rather than cladistics (Sokal and Sneath, 1963; Sneath and Sokal, 1973). Sometimes used in a more restricted sense as referring just to individuals or to populations.

Order A taxonomic category above **family** and below **class** (q.v.).

Ordered character (**Wagner character**) A character for which state transitions can occur in both directions and which if multistate, state transitions are additive.

Original description Generally the first publication that names a new taxon and attempts to characterize it such that it can be distinguished from other existing taxa (also see **protologue**).

Original designation The designation (specification; fixation) of the type species of a genus by an unambiguous statement in the original description.

OTU **Operational taxonomic unit** (q.v.).

Ouchterlony double diffusion An immunological procedure in which a polyclonal antibody and one or more antigen samples are allowed to diffuse into one another in a supporting medium (usually agarose gel). Precipitin arcs are formed where concentrations of antibody and corresponding antigen components reach appropriate concentrations (Ouchterlony, 1953; Maxson and Maxson, 1990; see chapter 9).

Outgroup A taxon or group of taxa believed to have a sister group relationship with or by paraphyletic with respect to a group of taxa under study (**ingroup** q.v.) and thus by the parsimony criterion, likely to display the plesiomorphous character state for the ingroup (Watrous and Wheeler, 1981; Maddison *et al.*, 1984; chapter 2).

Outgroup method A widely accepted method for determining the direction of character state change (character polarity) through comparison with the state(s) displayed by an appropriate outgroup (Watrous and Wheeler, 1981; Donoghue and Cantino, 1984; Maddison *et al.*, 1984).

PAGE (polyacrylamide gel electrophoresis) Any electrophoretic separation (usually of proteins) undertaken using a polyacrylamide gel as the supporting medium (cf. **SDS-PAGE**).

Panmixis A breeding system in which all members of a population or species are equally likely to cross-fertilize any other.

Paralectotype (also but less often **lectoparatype**) In zoological nomenclature the remaining specimens of a type series (i.e. **paratypes**) after a **lectotype** (q.v.) has been designated (Vorster, 1986). Botanical nomenclature through its provision for only one type specimen does not officially acknowledge the concept of the paralectotype.

Paralogous gene A gene whose homology with another gene results from the duplication of one of them.

Paraphyletic and **paraphyly** Several definitions of paraphyly have been proposed (Hennig, 1966; Nelson, 1971; Ashlock, 1971; Farris, 1974; Platnick, 1977a; Wiley, 1981). As used here it designates a group including a common ancestor and whose membership is defined by possession of a uniquely derived character state but one which may have undergone one or several reversals and therefore does not include all descendants of that common ancestor (Farris, 1974; Platnick, 1977a) (cf. **polyphyletic**).

Paratype Specimens of the type series other than the holotype. Sometimes used in a more restricted sense in botanical taxonomy (Faegri, 1983; Sivarajan, 1991).

Parsimony In reference to the widely employed cladistic principle that the most likely phylogenetic explanation must be the one requiring the least number of evolutionary steps (i.e. the most parsimonious explanation) (Friday, 1982; Farris, 1982a; Felsenstein, 1983; Fitch, 1971, 1984; Nordal, 1987; Sober, 1988).

Pathovar or **pathotype** In bacterial taxonomy an infrasubspecific rank of no official standing denoting a strain distinguishable on the basis of its pathogenicity spectrum.

Patristic Relating to similarity due to evolutionary descent (i.e. due to having a common ancestor).

Patronym A scientific name based on and in honour/recognition of a named person.

PCR **Polymerase chain reaction** (q.v.).

Peptides Typically short sequences of amino acids joined by peptide linkages (R–CO–NH–R') as in proteins.

Phage Any virus attacking bacteria. Often capable of causing lysis, (see also **lambda phage**).

Phagovar or **phagotype** In bacterial taxonomy an infrasubspecific rank of no official standing denoting a strain distinguishable on the basis of its susceptibility to certain phages.

Phenetics Now usually used to refer to **numerical taxonomy** (q.v.) and especially to methods that cluster taxa on the basis of similarity (Sneath and Sokal, 1973; see chapter 4).

Phenogram Any branching diagram with terminal taxa whose structure represents phenotypic or genotypic similarity between taxa rather than an evolutionary relationship (although the two may rarely be the same).

Phenon A word describing clusters of *OTU*s (q.v.) obtained from **phenetic** q.v. (numerical taxonomic) studies.

Phylogenetic systematics Another name for **cladism** (q.v.).

Phylogeny The evolutionary history of a group of taxa.

Phylogram A dendrogram indicating a hypothesized evolutionary history (i.e. one derived from synapomorphy data) which additionally indicates by means of branch length, the degree of evolutionary change believed to have occurred along each lineage.

Plasmid A (usually) circular, small DNA molecule separate from the main chromosome(s) but still replicated along with the main block of genetic material. Common in prokaryotes and widely employed in genetic manipulations (see also **cosmid**) (section 11.2.7).

Plenary powers The International Commissions of Botanical, Zoological and Bacterial Nomenclature reserve the right to make nomenclatural decisions in contradiction to the rules of their respective Codes if they deem it to be sufficiently desirable to do so, usually for the maintenance of stability where important taxa are involved.

Pleomorphic Of asomycete and basidiomycete fungi in which a species is represented by separate asexual and sexual spore-producing morphs (**anamorphs** and **teleomorphs**, respectively). The ICBN allows independent names for anamorphs until the teleomorph is known.

Plesiomorphy The ancestral character state for a character in a group of organisms (cf. **apomorphy**).

Plesion A special suprageneric category for fossil taxa which can effectively stand for, without specifying it, any higher taxonomic category (Patterson and Rosen, 1977; Wiley, 1981; Donoghue *et al.*, 1989). Not currently recognized by any of the Codes.

Polarity Refers to the evolutionary direction of a character state transition; a state presumed to have been displayed by an (common) ancestor then being coded as **plesiomorphic** (primitive) (q.v.) and other states **apomorphic** (advanced, derived) (q.v.) (see section 2.2.2).

Polyclade A multiple-entry identification key either using punched cards or implemented on a computer (see section 5.2.2).

Polyclonal antibody A mixture of antibodies produced against a single antigen by a population of lymphocytes (white blood cells) as, for example, when an antigen is inoculated into an appropriate animal (cf. **monoclonal antibody**).

Polymerase chain reaction or **PCR** A recently developed *in vitro* procedure for selectively amplifying (producing large numbers of identical copies of) small samples of DNA without the need for **cloning** (q.v.) (see chapter 11). Widely used where original material is in limited supply as with small organisms or fossil material (Brown, 1990; Innis *et al.*, 1990; McPherson *et al.*, 1991).

Polymorphism (1) The occurrence in one species of more than one discrete phenotypic state (either genetically based or otherwise); (2) the occurrence of two or more states of a character among the members of a taxon (Howard, 1988).

Polyphyletic and **polyphyly** Several definitions of polyphyly have been proposed (Hennig, 1966; Nelson, 1971; Ashlock, 1971; Farris, 1974). As used here it designates a group which does not include the common ancestor of all of its members (Farris, 1974; Platnick, 1977a) (cf. **paraphyletic**; see section 2.4.1).

Polyploidy   State when chromosome number is increased above normal diploid number ($2n$) initially by an integer multiple of n but this may be modified subsequently (see also **aneuploidy**, **euploidy**) (Jackson, 1976, 1982; Orr, 1990; Gastony, 1991; Thompson and Lumaret, 1992).

Polytene chromosomes   Chromosomes composed of many complete and aligned copied of each particular chromosome's DNA (chromatids). Such chromosomes are large and display distinct banding patterns which can be used to identify homologous regions. They occur in many larval Diptera (true flies), especially in salivary glands, some springtails (Insecta: Collembola) and in dinoflagellates and a few other Protista. Typically stained using orcein in acetic acid or examined *in vivo* using Nomarsky microscopy.

Polythetic   (1) Of a taxon whose membership is defined not by possession of a single characteristic, but by having a defined minimum number of a set of attributes; (2) of a key in which each lead of a couplet may make use of more than one character (cf. **monothetic**).

Polytomy   A branch point in a tree at which three or more branches arise from the ancestral line.

Precipitin reaction   An immunological reaction in which mixtures of antibody and antigen at appropriate concentrations form a precipitate (precipitin). Optimal precipitin formation occurs where the molecular proportion of antibody to antigen is approximately 1.5:1 (see section 9.2).

Primary homonym   Either of two identical species names that were both originally erected for (usually but not necessarily different) taxa in the same genus; only one can be a valid name, the other needing replacement, cf. **secondary homonym**.

Primer   A short section of single-stranded DNA that, when annealed to a longer complementary strand, enables DNA polymerase to extend the primer chain along the length of the longer DNA strand. Integral to the **polymerase chain reaction** (q.v.).

Principal components analysis   An ordination procedure widely used to determine how many discrete groups a set of **OTU**s (q.v.) belong to (section 4.4.1).

Principal coordinates analysis   An ordination procedure widely used to determine how many discrete groups a set of **OTU**s (q.v.) belong to (section 4.4.2).

Priority   A fundamental principle of some nomenclatural systems which establishes that the correct name of a taxon, if more than one name has been given, is the one which was published on the earliest date. Priority only extends back as far as a defined date and may be over-ruled for the sake of stability. Applies to botanical and zoological nomenclature.

Prokaryote   State displayed by organisms which lack a membrane bound nucleus and complex organelles, specifically the Monera.

Protologue   Collectively all information pertaining to a taxon or its type specimen(s) given in the original description. A term employed mostly in botanical nomenclature but useful elsewhere.

Pseudogene A gene which is no longer functional due to some change in its control region. Its selective neutrality makes it particularly attractive for phylogeny reconstruction (Joysey, 1988).

Punctuated equilibrium A model of Darwinian evolution in which changes result from periods of rapid evolution separated by periods of relative constancy.

Q-banding A simple chromosome banding technique visualized under ultraviolet light after quinacrine mustard staining. Fluorescent regions correspond to A–T rich regions although not all A–T rich regions react strongly (Macgregor and Varley, 1983).

RAPD or **random amplified polymorphic DNA** Genetic variation revealed by amplifying (see **PCR**) random DNA sequences using randomly chosen **primers** (q.v.) (Williams *et al.*, 1990; Welsh and McClelland, 1990; Chapco *et al.*, 1992; Hadrys *et al.*, 1992) (section 11.3.2).

R-banding Usually refers to a giemsa-based chromosomal banding procedure giving the reverse pattern to **G-banding** (q.v.) (Macgregor and Varley, 1983).

rDNA DNA sequences that code for ribosomal RNA and the associated spacer regions.

Regnum Latin for **kingdom** (q.v.), most often employed in botanical nomenclature.

Rejected name Any name for an organism other than its **valid name** (q.v.).

Rejected work A published work in which any nomenclatural acts are either unavailable under the relevant Code or which have been rejected (and therefore have no standing in nomenclature) by a specific decision of the relevant Commission.

Repetitive DNA DNA sequences represented by more than one copy in the haploid genome. Includes moderately repetitive sequences such as genes coding for tRNAs or rRNAs which may be present in tens or hundreds of copies, and highly repetitive DNA usually comprising short, non-coding tandemly repeated sequences that may be present in tens or hundreds of thousands of copies (see also **satellite DNA**).

Replacement name A scientific name proposed to replace a name that has been found to be a junior homonym (see section 6.4.3).

Restriction enzyme or **restriction endonuclease** An enzyme that cleaves double-stranded DNA, e.g. BamH1, EcoR1, Sau3I (see section 11.5).

Restriction fragment length polymorphism or **RFLP** The occurrence in a taxon of more than one morph defined by the presence or absence of a particular restriction enzyme recognition (cleaving) site (Templeton, 1983; Gillet, 1991).

Restriction site A short DNA nucleotide sequence (typically palindromic) recognized by Type II restriction enzymes which cleave the DNA adjacent to this site (Templeton, 1983).

Retention index   A measure, applicable to either a single character or a whole set of characters, that reflects for a given tree topology how much of the variation displayed by the character can be attributed to true synapomorphy.

Reticulate evolution   An evolutionary scenario in which hybridization brings about gene flow between clades and which can result in the creation of entirely new taxa (Humphries and Funk, 1984).

Revision   A taxonomic work which critically reappraises the classification of a group usually in the light of the discovery of additional material.

RFLP   See **restriction fragment length polymorphism**.

Robertsonian fusion   The fusion of two typically acrocentric chromosomes with the loss of one of the two centromeres.

Rooted tree   An hypothesized phylogeny to which the common ancestor has been attached.

SAHN   Sequential, agglomerative, hierarchic, non-overlapping clustering methods (see chapter 4).

Sanger–Coulson method   A method for sequencing DNA using DNA polymerase to synthesize a copy of the chain being sequenced and incorporating a small proportion of a dideoxynucleotide which will terminate chain elongation where that nucleotide occurs in the sequence (see chapter 11), cf. **Maxam–Gilbert method**.

Satellite DNA   Highly repeated (often more than 100 000 copies per genome) tandemly arranged DNA sequences (usually less than 1000 base pairs) (see Macgregor and Sessions, 1986).

Scientific name   For species, its binomen (i.e. its genus and species name), for a subspecies, its trinomen (i.e. its genus, species and subspecies name).

Schizotype   An implied lectotype.

SDS   **Sodium dodecylsulphate** (q.v.).

SDS–PAGE   **PAGE** (q.v.) in which **SDS** (q.v.) has been incorporated into the gel in order to enable proteins to be separated electrophoretically in accordance with their approximate molecular weights (see section 10.2.1).

Secondary homonym   Species level names that although originally described under separate genera became homonyms when they were subsequently included under the same generic name by a revising author (cf. **primary homonym**).

Section   An informal taxonomic rank between genus and series used principally in botanical and bacteriological nomenclature (Rasnitsyn, 1982).

Segment coding   A method for discretely coding a continuously variable character (chapter 2; Thorpe, 1984; Farris, 1990).

Semantide   A character that directly reflects the information contained within an organism's genome, such as DNA, RNA or protein sequences or such of their attributes that can be related directly to the DNA sequence, such as the electrophoretic mobility of a protein or its immunological determinants but not its enzymic function.

Series   (1) An informal supplementary taxonomic rank between section and species used principally in botanical and bacteriological nomenclature (Rasnitsyn, 1982); (2) a number of specimens collected at the same time and locality by a collector (see also **type series**).

Serovar or **serotype (1)**   In bacterial taxonomy an infrasubspecific rank of no official standing denoting a strain distinguishable on the basis of its antigenic properties.

Serotype (2)   In general taxonomy, a group of **OTU**s characterized by a particular immunological feature.

Sibling species   Either of two very closely related species which can only be distinguished by minute or cryptic characters but are genetically isolated from one another. By implication, these are **sister groups** (q.v.) (Mayr, 1942).

Similarity   A measure of the similarity between two taxa or **OTU**s (q.v.) that increases with increasing similarity (cf. distance; see also **genetic distance**) (Gower, 1971; Sneath and Sokal, 1973).

Single linkage clustering   An agglomerative clustering technique based on the minimum distance between an OTU and a cluster of OTUs to which it is to be joined.

Sister group   Either of a pair of taxa or groups whose closest common ancestor is not shared by any other group. When given formal taxonomic status both should be accorded the same rank. The choice of gender is in fact an accident of German grammar and it is worth noting that the French equivalent is 'groupe-frère'.

Sodium dodecylsulphate   A detergent which binds to proteins enabling them to be separated electrophoretically according to their molecular weights. Usually abbreviated 'SDS'; see also **SDS-PAGE**.

Soredia   Asexually produced propagule produced by some lichen clones comprising both a fungal and an algal component (Tehler, 1982).

Southern blot   A method of revealing lengths of DNA fragments by transferring denatured fragments from electrophoresis gels, in which they have been previously separated according to size, on to nitrocellulose paper and then identifying the positions of selected sequences on the paper by binding radio-labelled, complementary DNA sequences and autoradiographing the results.

sp.   Abbreviation for 'species' (singular). Used when true identity unknown, e.g. *Pieris* sp.

spp.   Abbreviation for 'species' (plural). Used to indicate several species usually of the same genus.

Species   A taxonomic level subject to various definitions; see **biological species**, phylogenetic species.

Species-group name   Specific and subspecific names.

Specific name   Technically the combination of a generic name and a trivial name although commonly now applied to just the trivial name in a binomen (cf. **epithet**).

Splitter   An informal description of a taxonomist who tends to divide taxa on the basis of small differences (cf. **lumper**).

ssp.   Abbreviation for 'subspecies'.

Steiner tree   The shortest tree connecting a series of points on a plain. Hence the hypothetical DNA sequence from which a series of observed sequences could be obtained by the minimum number of evolutionary changes.

Strain   In bacteriology, a pure culture of the descendents of a single isolation from the 'wild'.

Subjective synonym   Synonymy resulting from the opinion of a taxonomist that two or more taxa with different names are in fact the same (cf. **objective synonym**).

Sympatry   The occurrence together at the same locality or in overlapping areas of two populations.

Symplesiomorphy   **Plesiomorphous** character (q.v.) states shared by a group of taxa due to shared ancestry.

Synapomorphy   An **apomorphous** character (q.v.) shared by two or more taxa and thus indicating common ancestry for the members of this group.

Synonym   Each of a set of different generic or specific names that can be applied to a single taxon usually as a result of the same taxon having been described on more than one separate occasion (usually but not always by different authors) (see also **subjective synonym** and **objective synonym**).

Synoptic collection   A collection of correctly identified representatives of all taxa from a given area used as a reference collection to help confirm subsequent identifications.

Syntype   Any member of a type series for which no holotype or lectotype has been designated (Faegri, 1983).

Systematics   A word given a range of definitions but generally held to be more or less equivalent to taxonomy. Here it is treated in a similar way to that described by Hawksworth and Bisby (1988), i.e. as somewhat broader than taxonomy and including in its covenance aspects of phylogeny, evolution, biogeography and genetics.

Tautonymy   The identical (or sometimes virtually identical) spelling of a generic, specific or subspecific name. Of relevance in subsequent fixation of a type species, etc.

Taxon   Any definable taxonomic unit whether described or not, e.g. subspecies, species, tribe, genus, family, etc. See also **operational taxonomic unit**.

Taxonomy   As with **systematics** (q.v.) a word denoting slightly different things to different people. Here taken to be a subsection of systematics that deals with **classification** (q.v.), **nomenclature** (q.v.) and other aspects of defining, naming and identifying taxa.

Telocentric   Of a chromosome with the centromere located at the extreme end of its single arm. These are rather rare and are probably evolutionarily unstable.

Topotype A specimen originating from the type locality of a species and believed to belong to that species.

Tracer In **DNA hybridization** (q.v.) the sheared pieces of radio-labelled DNA which bind to the **driver DNA** (q.v.) (see section 11.6).

Transformation series An hypothesized set of allowed (likely) transitions between multiple character states (Mickevich, 1982; for a chemotaxonomic example see Seaman and Funk, 1983).

Transition A nucleotide substitution in which a purine is replaced by another purine or a pyrimidine by another pyrimidine (cf. **transversion**).

Translocation A karyotype rearrangement in which a segment of one chromosome becomes inserted somewhere else in the genome.

Transversion A nucleotide substitution in which a purine is replaced by a pyrimidine or vice versa (cf. **transition**).

Tribe A formal classificatory group between family and genus (also stem of subtribe and supertribe).

Trinomen A scientific name comprising a genus, species and subspecies name.

True character A character whose state distribution among taxa correctly reflects the evolutionary history of the group, i.e. all taxa possessing a given apomorphous character state comprise a monophyletic group and all those lacking it do not belong to the group, cf. **false character** (see section 3.1.2).

Type See **type species**, **type specimen** and **co-**, **holo-**, **iso-**, **lecto-**, **meta-**, **neo-**, **para-**, **paralecto-**, **syn-**, and **topotype**.

Type culture A living pure culture derived from a single isolation of a prokaryote being the effective type material upon which the prokaryote species are based. Incorrectly used for Protista, Fungi or other multicellular organisms for which there should be a permanently preserved holotype although the lodging of a culture derived from the holotype in a reputable culture collection is highly recommended.

Type series All specimens of a species upon which the original author based the original description but excluding any that the author specifically mentions as variants or as doubtful members of the species.

Type slide In zoological nomenclature a special form of type specimen consisting of a series of directly related individuals of a protozoan mounted as a microscopic preparation and used to define the species concept. No one individual in such a type series can be regarded as either a **holotype** or **lectotype**.

Type species The species designated as the type of a **genus-group name**.

Type specimen The specimen or any member of a series of specimens (type series) on which the original description and hence the concept of the species is fixed.

UAR (**uniform average rate**) The concept, now widely refuted, that the rate of accumulation of mutations in DNA or a protein for example, will be more or less constant on average along all evolutionary branches. The UAR concept is therefore relevant to the idea that the age of separation of lineages can be dated by reference to the number of mutations accumulated.

Uninformative character A character whose variation among a set of taxa provides no information about their phylogenetic relationships, e.g. **autapomorphies** (q.v.).

Unordered character A multistate character for which transitions between any two states are equally likely (or unlikely).

Unrooted trees A network connecting a set of taxa which has not been connected to the common ancestor.

UPGMA (unweighted pair-group method using arithmetic averages) A clustering procedure employed widely in numerical taxonomy (Sneath and Sokal, 1973; section 4.3.3). No longer recommended for most taxonomic purposes especially for phylogeny reconstruction.

Valid name The correct scientific name for a taxon (cf. **rejected name, invalid name**).

Variety A supplementary taxonomic category between **subspecies** and **form** (q.v.) in botanical nomenclature. Not accepted for taxonomic purposes by the **ICZN**.

Vector A virus (i.e. a **phage** q.v.) or circular DNA molecule used as a means of introducing a sequence of DNA from a **donor** (q.v.) organism into a bacterium for cloning (q.v.) purposes, i.e. in order to amplify the number of copies of the donor DNA.

Vicariance Pertaining to the presence of closely related taxa in different geographic areas as a result of the formation of natural barriers dividing an existing population as opposed to through jump dispersal (Croizat *et al.*, 1974; Nelson and Platnick, 1981; Cracraft, 1982; Humphries and Parenti, 1986; Myers and Giller, 1988).

Voucher specimen A specimen (or its remains) that was used in a published study and which is deposited in a permanent collection so as to provide evidence regarding the identity of the material studied should there ever become a need to check the original identification (Francoeur, 1976; Yoshimoto, 1978; see section 13.1.5).

Wagner character **Ordered characters** (q.v.) as used by the Wagner parsimony method.

Wagner ground plan divergence analysis A method for obtaining phylogenetic hypotheses based on the total numbers of shared synapomorphies displayed by pairs of taxa (Wagner, 1961, 1980; Wiley, 1981).

Weighting The process of assigning differential importance to different character systems usually employed either in phylogenetic analysis or identification systems (e.g. keys). Weighting may be decided arbitrarily, on the basis of empirical knowledge (e.g. about relative frequencies of **transitions** and **transversions** q.v.), on the basis of compatibility with other characters or length of fit on a tree. See also *a posteriori* and *a priori* **weighting**.

WGPA or **WGDA**   **Wagner ground plan analysis** (q.v.).

WPGMA (weighted pair-group method using arithmetic averages)   A phenetic clustering technique similar to **UPGMA** (Sneath and Sokal, 1973; section 4.3.4).

Zymogram   An electrophoretic gel in which the positions of zones of activity of a particular enzyme have been revealed by specific staining (see section 10.3).

BIBLIOGRAPHY

This bibliography includes mostly references referred to specifically in the text, although a few other publications that have not been specifically mentioned but nevertheless include interesting discussions on taxonomy are also listed.

Abbott, L. A., Bisby, F. A. and Rogers, D. J. (1985) *Taxonomic Analysis in Biology*, Columbia University Press, New York.

Ackermann, H.-W. and Eisenstark, A. (1974) The present state of phage taxonomy. *Intervirology*, **3**, 201–219.

Adams, E. N. (1972) Consensus techniques and the comparison of taxonomic trees. *Syst. Zool.*, **21**, 390–397.

Adams, R. P. (1982) A comparison of multivariate methods for the detection of hybridization. *Taxon*, **31**, 646–661.

Albert, V. A., Williams, S. E. and Chase, M. W. (1992) Carnivorous plants: Phylogeny and structural evolution. *Science*, **257**, 1491–1495.

Alderson, G. (1985) The application and relevance of non-hierarchic methods in bacterial taxonomy. In *Computer-assisted Bacterial Systematics*, M. Goodfellow, D. Jones and F. G. Priest (eds), Academic Press, London, pp. 227–263.

Anderson, G. J. and Levine, D. A. (1982) Three taxa constitute the sexes of a single dioecious species of *Solanum*. *Taxon*, **31**, 667–672.

Archie, J. W. (1985) Methods for coding variable morphological features for numerical taxonomic analysis. *Syst. Zool.*, **34**, 236–345.

Archie, J. W. (1989a) A randomization test for phylogenetic information in systematic data. *Syst. Zool.*, **38**, 239–252.

Archie, J. W. (1989b) Homoplasy excess ratios: new indicates for measuring levels of homoplasy in phylogenetic systematics and a critique of the consistency index. *Syst. Zool.*, **38**, 253–269.

Arnason, U. and Best, P. B. (1991) Phylogenetic relationships within the Mysticeti (whalebone whales) based upon studies of highly repetitive DNA in all extact species. *Hereditas*, **114**, 263–269.

Arnold, E. N. (1981) Estimating phylogenies at low taxonomic levels. *Z. Zool. Syst. Evolut.-forsch.*, **19**, 1–35.

Ashlock, P. D. (1971) Monophyly and associated terms. *Syst. Zool.*, **20**, 63–69.

Ashlock, P. D. (1974) The uses of cladistics. *Ann. Rev. Ecol. Syst.*, **5**, 81–99.

Atkinson, W. D. and Gammerman, A. (1987) An application of expert systems technology to biological identification. *Taxon*, **36**, 705–714.

Austin, B. and Priest, F. (1986) *Modern Bacterial Taxonomy*, Van Nostrand Reinhold (UK), Wokingham.

Avise, J. C. (1975) Systematic value of electrophoretic data. *Syst. Zool.*, **23**, 465–481.

Avise, J. C. and Aquandro, C. F. (1982) A comparative summary of genetic distances. *Evol. Biol.*, **15**, 151–185.

Avise, J. C., Arnold, J., Ball, R. M., Bermingham, E., Lamb, T., Neigel, J. E., Reeb, C. A. and Saunders, N. C. (1987) Intraspecific phylogeography: the mitochondrial

DNA bridge between population genetics and systematics. *Ann. Rev. Ecol. Syst.*, **18**, 489–522.

Ayala, F. J. (ed.) (1976) *Molecular Evolution*, Sinauer Associates, Sunderland, Massachusetts.

Ayensu, E. S. (1984) Taxonomic problems relating to endangered plant species. In *Current Concepts in Plant Taxonomy. Systematics Association Special Volume 25*, V. H. Heywood and D. M. Moore (eds), Academic Press, London, chapter 20, pp. 411–421.

Aynilian, G. H., Farnsworth, N. R. and Trojanek, T. (1973) The use of alkaloids in determining taxonomic position of *Vinca libanotica* (Apocynaceae). In *Chemistry in Botanical Classification*, (G. Bendz and J. Santesson (eds), New York, pp. 189–204.

Ball, I. R. (1976) Nature and formulation of biogeographical hypotheses. *Syst. Zool*, **24**, 407–430.

Barigozzi, I. (ed.) (1982) *Mechanisms of Speciation. Volume 96, In Progress in Clinical and Biological Research*, Alan R. Liss, New York.

Baum, B. R. (1976) Weighting character states. *Taxon*, **25**, 257–260.

Baum, B. R. (1988) A simple procedure for establishing discrete characters from measurement data, applicable to cladistics. *Taxon*, **37**, 63–70.

Baum, D. (1992) Phylogenetic species concepts. *Trends Ecol. Evol.*, **7**, 1–2.

Behnke, H.-D. (1977) Transmission electron microscopy and systematics of flowering plants. *Pl. Syst. Evol. Suppl.*, **1**, 381–400.

Benirschke, K. and Kumamoto, A. T. (1991) Mammalian cytogenetics and conservation of species. *J. Hered.*, **82**, 187–191.

Bennett, S. T., Kenton, A. Y. and Bennett, M. D. (1992) Genomic *in situ* hybridization reveals the allopolyploid nature of *Milium montianum* (Gramineae). *Chromosoma*, **101**, 420–424.

Blackwelder, R. E. (1967) *Taxonomy, a Text and Reference Book*. John Wiley & Sons, New York.

Borst, P., Tabak, H. F. and Grivell, L. A. (1983) Extranuclear genes. In *Eukaryotic Genes. Their Structure, Activity and Regulation*, N. Maclean, S. P. Gregory and R. A. Flavell (eds), Butterworths, London, pp. 71–84.

Bramwell, D. (1989) Taxonomy: sands of time. *Taxon*, **38**, 404–405.

Bremer, K. (1990) Combinable component consensus. *Cladistics*, **6**, 369–372.

Bremer, K. and Wanntorp, H.-E. (1978a) Phylogenetic systematics in botany. *Taxon*, **27**, 317–329.

Bremer, K. and Wanntorp, H.-E. (1978b) Geographic populations or biological species in phylogeny reconstruction. *Syst. Zool.*, **28**, 220–224.

Brooks, D. R., Mayden, R. L. and McLennan, D. A. (1992) Phylogeny and biodiversity: conserving our evolutionary legacy. *Trends Ecol. Evol.*, **7**, 55–59.

Brooks, D. R. and McLennan, D. A. (1991) *Phylogeny, Ecology and Behaviour: A Research Program in Comparative Biology*, University of Chicago Press, Chicago and London.

Brown, V. K. (1983) *Grasshoppers*, Cambridge University Press, Cambridge.

Brown, W. M. (1985) The mitochondrial genome of animals. In *Molecular Evolutionary Genetics*, R. J. MacIntyre (ed.), Plenum Press, New York, pp. 95–129.

Brown, T. A. (1990) *Gene cloning. An Introduction*, Chapman & Hall, London.

Bruce, E. J. and Ayala, F. J. (1979) Phylogenetic relationship between man and the apes: electrophoretic evidence. *Evolution*, **33**, 1040–1056.

Bruns, T. D., White, T. J. and Taylor, J. W. (1991) Fungal molecular systematics. *Ann. Rev. Ecol. Syst.*, **22**, 525–564.

Buchanan, R. E. and Gibbons, N. E. (1974) *Bergey's Manual of Determinative Bacteriology (8th edn)*, Williams and Wilkins, Baltimore.

Camin, J. H. and Sokal, R. R. (1965) A method for deducing branching sequences in phylogeny. *Evolution*, **19**, 311–326.

Cann, R. L., Stoneking, M. and Wilson, A. C. (1987) Mitochondrial DNA and human evolution. *Nature*, **325**, 31–36.

Cannon, P. F. (1986) International Commission on the Taxonomy of Fungi (ICTF): name changes in fungi of microbiological, industrial and medical importance. Parts 1–2. *Microbiol. Sci.*, **29**, 168–171, 285–287.

Cano, R. J., Poinar, H. and Poinar, G. O., Jr (1992) Isolation and partial characterization of DNA from *Proplebeia dominicana* (Hymenoplera: Apidae) in 25–40 million year old amber. *Med. Sci. Res.*, **20**, 249–251.

Carpenter, J. M. (1988) Choosing among multiple equally parsimonious cladograms. *Cladistics*, **4**, 291–296.

Cavalier-Smith, T. (1987) Eukaryotes with no mitochondria. *Nature*, **326**, 332–333.

Cavalli-Sforza, L. L. and Edwards, A. W. F. (1967) Phylogenetic analysis. Models and estimation procedures. *Evolution*, **21**, 550–570.

Ceska, A. and Trampour, A. D. (1979) Computer editing of serial and indented identification keys. *Taxon*, **28**, 329–335.

Chalmers, N. R. (1992) The role of museums. In *Taxonomy in the 1990's*, Linnean Society of London/The Systematics Association, pp. 29–30.

Champion, A. B., Prager, E. M., Watcher, D. and Wilson, A. C. (1974) Microcomplement fixation. In *Biochemical and Immunological Taxonomy of Animals*, C. A. Wright (ed.), Academic Press, New York, pp. 397–416.

Chapco, W., Ashton, N. W., Martel, R. K. B., Antonishyn, N. and Crosby, W. L. (1992) A feasibility study of the use of random amplified polymorphic DNA in the population genetics and systematics of grasshoppers. *Genome*, **35**, 569–574.

Charig, A. J. (1982) Systematics in biology: a fundamental comparison of some major schools of thought. In *Problems of Phylogenetic Reconstruction*, K. A. Joysey and A. E. Friday (eds), Academic Press, London, pp. 363–430.

Claridge, M. F. (1988) Species concepts and speciation in parasites. In *Prospects in Systematics. Systematics Association Special Volume No. 36*, D. L. Hawksworth (ed.), Clarendon Press, Oxford, pp. 92–111.

Claridge, M. F. (1991) Biological diversity and systematic biology: the need for stable nomenclature. *Biologist*, **38**, 103–104.

Claridge, M. F. and Ingrouille, M. (1992) Systematic biology and higher education in the U.K. In *Taxonomy in the 1990's*, The Linnean Society of London and The Systematics Association, London, pp. 39–48.

Clarkson, S. G. (1983) Transfer RNA genes. In *Eukaryotic Genes. Their Structure, Activity and Regulation*, N. Maclean, S. P. Gregory and R. A. Flavell (eds), Butterworths, London, chapter 14, pp. 239–261.

Clifford, H. T. and Stephenson, W. (1975) *An Introduction to Numerical Classification*, Academic Press, New York.

Clifford, H. T., Rogers, R. W. and Dettmann, M. E. (1990) Where now for taxonomy? *Nature*, **346**, 602.

Coats, S. A., Wicker, L. and McCoy, C. W. (1990) Protein variation among Fuller rose beetle populations from Florida, California, and Arizona (Coleoptera: Curculilonidae). *Ann. Entomol. Soc. Amer.*, **83**, 1054–1062.

Coddington, J. A. and Levi, H. W. (1991) Systematics and evolution of spiders (Araneae). *Ann. Rev. Ecol. Syst.*, **22**, 565–592.

Cole, G. T. (1979) Contributions of electron microscopy to fungal classification. *Amer. Zool.*, **19**, 589–608.

Colless, D. H. (1980) Congruence between morphometric and allozyme data for *Menidia* species: a reappraisal. *Syst. Zool.*, **29**, 288–299.

Comings, D. E. (1978) Mechanisms of chromosome banding and implications for chromosome structure. *Ann. Rev. Genet.*, **12**, 25–46.

Cotgreave, P. and Harvey, P. H. (1992) Relationship between body size, abundance and phylogeny in bird communities. *Funct. Ecol.*, **6**, 248–256.

Cowan, R. S. and Stafleu, F. A. (1982) The origins and early history of I.A.P.T. *Taxon*, **31**, 415–420.

Cracraft, J. (1982) Geographic differentiation, cladistics and vicariance biogeography: reconstructing the tempo and mode of evolution. *Amer. Nat.*, **22**, 411–424.

Cracraft, J. (1983) Species concepts and speciation analysis. *Curr. Ornithol.*, **1**, 159–187.

Cracraft, J. (1987) DNA hybridization and avian phylogenies. In *Evolutionary Biology. Volume 21*, M. K. Hecht, B. Wallace and G. T. Prance (eds), Plenum Press, New York, chapter 7, pp. 179–235.

Cracraft, J. (1989) Speciation and its ontogeny: The empirical consequences of alternative species concepts for understanding patterns and processes of differentiation. In *Speciation and its Consequences*, D. Otte and J. A. Endler (eds), Sinauer Associates, Sunderland, Massachusetts, pp. 28–59.

Crawford, D. J. (1978) Flavonoid chemistry and angiosperm evolution. *Bot. Rev.*, **44**, 431–456.

Crawford, D. J. (1990) *Plant Molecular Systematics. Macromolecular Approaches*, John Wiley & Sons, New York.

Crisci, J. V. (1984) Taxonomic congruence. *Taxon*, **33**, 233–239.

Crisci, J. V. and Stuessy, T. F. (1980) Determining primitive character states for phylogenetic reconstruction. *Syst. Bot.*, **5**, 112–135.

Croizat, L., Nelson, G. and Rosen, D. E. (1974) Centers of origin and related concepts. *Syst. Zool.*, **23**, 265–287.

Crowson, R. A. (1970) *Classification and Biology*, Heinemann Educational Books, London.

Czaja, A. T. (1978) Structure of starch grains and the classification of vascular plant families. *Taxon*, **27**, 463–470.

Dallwitz, M. J. (1974) A flexible computer program for generating identification keys. *Syst. Zool.*, **23**, 50–57.

Dallwitz, M. J. (1980) A general system for coding taxonomic descriptions. *Taxon*, **29**, 41–46.

Dallwitz, M. J. and Paine, T. A. (1986) *User's Guide to the DELTA System. A General System for Processing Taxonomic Descriptions (3rd edn)*, CSIRO, Division of Entomology Report No. 13, CSIRO, Canberra, Australia.

Daugherty, C. H., Cree, A., Hay, J. M. and Thompson, M. B. (1990) Neglected taxonomy and continuing extinction of tuatara (*Sphenodon*). *Nature*, **347**, 177–179.

Davison, D. (1985) Sequence similarity ('homology') searching for molecular biologists. *Bull. Math. Biol.*, **47**, 437–474.

DeBry, R. W. and Slade, N. A. (1985) Cladistic analysis of restriction endonuclease cleavage maps within a maximum-likelihood framework. *Syst. Zool.*, **34**, 21–34.

De Jong, R. (1980) Some tools for evolutionary and phylogenetic studies. *Z. Zool. Syst. Evolut.-forsch.*, **18**, 1–23.

De Queiroz, A. and Wimberger, P. H. (1993) The usefulness of behaviour for phylogeny estimation: levels of homoplasy in behavioural and morphological characters. *Evolution*, **47**, 46–60.

De Queiroz, K. and Donoghue, M. J. (1988) Phylogenetic systematics and the species problem. *Cladistics*, **4**, 317–338.

De Queiroz, K. and Donoghue, M. J. (1990) Phylogenetic systematics and species revisited. *Cladistics*, **6**, 83–90.

De Salle, R. and Grimaldi, D. A. (1991) Morphological and molecular systematics of the Drosophilidae. *Ann. Rev. Ecol. Syst.*, **22**, 447–475.

Diamond, J. M. (1988) DNA-based phylogenies of the three chimpanzees. *Nature*, **332**, 685–686.

Diamond, J. M. (1990) Old dead rats are valuable. *Nature*, **347**, 334–335.

Dixon, B. and Pohajdak, W. (1992) Did the ancestral globin gene of plants and animals contain only two introns? *Trends Biochem. Sci.*, **17**, 486–488.

Donoghue, M. J. (1985) A critique of the biological species concept and recommendations for a phylogenetic alternative. *Bryologist*, **88**, 172–181.

Donoghue, M. J. and Cantino, P. D. (1984) The logic and limitations of the outgroup substitution approach to cladistic analysis. *Syst. Bot.*, **9**, 192–202.

Donoghue, M. J., Doyle, J. A., Gauthier, J., Kluge, A. G. and Rowe, T. (1989) The importance of fossils in phylogeny reconstruction. *Ann. Rev. Ecol. Syst.*, **20**, 431–460.

Dubois, A. (1988) Le genre en zoologie: Essai de systématique théorique. *Mem. Mus. Natn. Hist. Nat. (A)*, **139**, 1–132.

Duckett, J. G. (1988) Electron microscopy in systematics: genesis and revelations. In *Prospects in Systematics. Systematics Association Special Volume No. 36*, D. L. Hawksworth (ed.), Clarendon Press, Oxford, pp. 217–233.

Dunbar, R. W. (1966) Cytotaxonomic studies in black flies (Diptera: Simulidae). *Chromosomes Today*, **1**, 179–181.

Duncan, T. (1984) Willi Hennig, character compatibility, Wagner parsimony, and the "Dendrogrammaceae" revisited. *Taxon*, **33**, 698–704.

Duncan, T. and Meacham, C. A. (1986) Multiple-entry keys for the identification of angiosperm families using a microcomputer. *Taxon*, **35**, 492–494.

Duncan, T. and Stuessy, T. F. (eds) (1984) *Cladistics: Perspectives on the Reconstruction of Evolutionary History*, Columbia University Press, New York.

Duncan, T., Phillips, T. R. and Wagner, W. H. Jr. (1980) A comparison of branching diagrams derived by various phenetic and cladistic methods. *Syst. Bot.*, **5**, 264–293.

Dunn, G. and Everitt, B. S. (1982) *An Introduction to Mathematical Taxonomy*, Cambridge University Press, Cambridge, UK.

Dupuis, C. (1984) Willi Hennig's impact on taxonomic thought. *Ann. Rev. Ecol. Syst.*, **15**, 1–24.

Edwards, A. N. F. and Cavalli-Sforza, L. L. (1964) Reconstruction of evolutionary trees. In *Phenetic and Phylogenetic Classification*, V. H. Heywood and J. McNeill (eds), Systematics Association, London, pp. 67–76.

Eggleton, P. (1991) Patterns in the male mating strategy of the Rhyssini: a holophyletic group of parasitoid wasps (Hymenoptera: Ichneumonidae). *Anim. Behav.*, **41**, 829–838.

Eldredge, N. and Cracraft, J. (1980) *Phylogenetic Patterns and the Evolutionary Process*, Columbia University Press, New York.

Eldredge, N. and Gould, S. J. (1972) Punctuated equilibria: an alternative to phyletic gradualism. In *Models in Palaeobiology*, T. J. M. Schopf (ed.), Freeman & Cooper, San Francisco, pp. 82–115.

Erdtmann, G. (1954) Pollen morphology and plant taxonomy. *Bot. Notiser*, 65–81.

Ereshefsky, M. (ed.) (1992) *The Units of Selection: Essays on the Nature of Species*, MIT Press, Massachusetts.

Erwin, T. L. (1982) Tropical forests: their richness in Coleoptera and other arthropod species. *Coleopt. Bull.*, **36**, 74–75.

Erzinclioglu, Y. Z. and Unwin, D. M. (1986) The stability of zoological nomenclature. *Nature*, **320**, 687.

Estabrook, G. F. (1972) Cladistic methodology: a discussion of the theoretical basis for the induction of evolutionary history. *Ann. Rev. Ecol. Syst.*, **3**, 427–456.

Estabrook, G. F. (1977) Does common equal primitive? *Syst. Bot.*, **2**, 36–42.

Estabrook, G. F., Stranch, J. G. and Fiala, K. L. (1977) An application of compatibility analysis to the Blackith's data on orthopteroid insects. *Syst. Zool.*, **26**, 269–276.

Faegri, K. (1983) The hierarchy of types. *Taxon*, **32**, 640–641.

Faith, D. P. (1992) Conservation evaluation and phylogenetic diversity. *Biol. Conserv.*, **61**, 1–10.

Faith, D. P. and Cranston, P. S. (1991) Could a cladogram this short have arisen by chance alone?: on permutation tests for cladistic structure. *Cladistics*, **7**, 1–28.

Farnsworth, N. R. (1988) Screening plants for new pharmaceuticals. In *Biodiversity*, E. O. Wilson (ed.), National Academy Press, Washington D.C., pp. 83–97.

Farris, J. S. (1969) A successive approximations approach to character weighting. *Syst. Zool.*, **18**, 374–385.

Farris, J. S. (1970) Methods for computing Wagner trees. *Syst. Zool.*, **19**, 83–92.

Farris, J. S. (1972) Estimating phylogenetic trees from distance matrices. *Amer. Nat.*, **106**, 645–668.

Farris, J. S. (1974) Formal definitions of paraphyly and polyphyly. *Syst. Zool.*, **23**, 548–554.

Farris, J. S. (1977) Phylogenetic analysis under Dollo's Law. *Syst. Zool.*, **26**, 77–88.

Farris, J. S. (1978) Inferring phylogenetic trees from chromosome inversion data. *Syst. Zool.*, **27**, 275–284.

Farris, J. S. (1981) Distance data in phylogenetic analysis. In *Advances in Cladistics. Proceedings of the First Meeting of the Willi Hennig Society*, V. A. Funk and D. R. Brooks (eds), New York Botanical Gardens, Bronx, pp. 3–23.

Farris, J. S. (1982a) Outgroups and parsimony. *Syst. Zool.*, **31**, 328–334.

Farris, J. S. (1982b) *PHYSIS*; an interactive package for assessing phylogenetic relationships, State University of New York, Stoney Brook.

Farris, J. S. (1983) The logical basis of phylogenetic systematics. In *Advances in Cladistics, Volume 2. Proceedings of the Second Meeting of the Willi Hennig Society*, N. I. Platnick and V. A. Funk (eds), Columbia University Press, New York, pp. 7–36.

Farris, J. A. (1986) Synapomorphy, parsimony, and evidence. *Taxon*, **35**, 298–315.

Farris, J. S. (1988) *Hennig86, version 1.5*. Computer program distributed by J. S. Farris, Port Jefferson Station, New York.

Farris, J. S. (1989) The retention index and the rescaled consistency index. *Cladistics*, **5**, 417–419.

Farris, J. S. (1990) Phenetics in camouflage. *Cladistics*, **6**, 91–100.

Feldmann, R. M. and Manning, R. B. (1992) Crisis in systematic biology in the 'age of biodiversity'. *J. Paleontol.*, **66**, 157–158.

Felsenstein, J. (1978a) The number of evolutionary trees. *Syst. Zool.*, **27**, 27–33.

Felsenstein, J. (1978b) Cases in which parsimony or compatibility methods will be positively misleading. *Syst. Zool.*, **27**, 401–410.

Felsenstein, J. (1981) Evolutionary trees from DNA sequences: a maximum likelihood approach. *J. Mol. Evol.*, **17**, 368–376.

Felsenstein, J. (1982a) Numerical methods for inferring evolutionary trees. *Q. Rev. Biol.*, **57**, 379–404.

Felsenstein, J. (1982b) How can we infer geography and history from gene frequencies? *J. Theor. Biol.*, **96**, 9–20.

Felsenstein, J. (1983) Parsimony in systematics: biological and statistical issues. *Ann. Rev. Ecol. Syst.*, **14**, 313–333.

Felsenstein, J. (1984) The statistical approach to inferring evolutionary trees and what it tells us about parsimony and compatibility. In *Cladistics: Perspectives on the Reconstruction of Evolutionary History*, T. Duncan and T. F. Stuessy (eds), Columbia University Press, New York, pp. 169–191.

Felsenstein, J. (1985a) Phylogenies and the comparative method. *Amer. Nat.*, **125**, 1–15.

Felsenstein, J. (1985b) Confidence limits on phylogenies: An approach using the bootstrap. *Evolution*, **39**, 783–791.

Felsenstein, J. (1987) *PHYLIP* – Phylogeny Inference Package; version 3.0. Documentation.

Felsenstein, J. (1988) The detecting of phylogeny. In *Prospects in Systematics. Systematics Association Special Volume No. 36*, D. L. Hawksworth (ed.), Clarendon Press, Oxford, pp. 112–127.

Fenner, F. (1976) Classification and nomenclature of viruses: Second report of the International Committee on Taxonomy of Viruses. *Intervirology*, **7**, 1–116.

Ferguson, A. (1980) *Biochemical Systematic and Evolution*, Blackie, Glasgow.

Ferguson, A. (1988) Isozyme studies and their interpretation. In *Prospects in Systematics*, D. L. Hawksworth (ed.), Clarendon Press, Oxford, pp. 184–201.

Fiala, K. L. and Sokal, R. R. (1985) Factors determining the accuracy of cladogram estimation: Evaluation using computer simulation. *Evolution*, **39**, 609–622.

Fitch, W. M. (1971) Toward defining the course of evolution: minimum change for a specific tree topology. *Syst. Zool.*, **20**, 404–416.

Fitch, W. M. (1984) Cladistic and other methods: Problems, pitfalls, and potentials. In *Cladistics: Perspectives on the Reconstruction of Evolutionary History*, T. Duncan and T. F. Stuessy (eds), Columbia University Press, New York, pp. 221–252.

Fitch, W. M. and Margoliash, E. (1967) Construction of phylogenetic trees. *Science*, **155**, 279–284.

Flowerdew, M. W. and Crisp, D. J. (1976) Allelic esterase isozymes, their variation with season, position on the shore and stage of development in the cirripede *Balanus balanoides*. *Mar. Biol.*, **35**, 319–325.

Ford, E. B. (1957) *Butterflies*, Collins, London.

Fortey, P. L. and Jefferies, R. P. S. (1982) Fossils and phylogeny – a compromise approach. In *Problems of Phylogenetic Reconstruction*, K. A. Joysey and A. E. Friday (eds), Academic Press, London, pp. 197–234.

Francoeur, A. (1976) The need for voucher specimens in behavioural and ecological studies. *Bull. Can. Entomol. Soc.*, **8**, 23.

Friday, A. E. (1982) Parsimony, simplicity, and what actually happened. *Zool. J. Linn. Soc.*, **74**, 329–335.

Friedmann, E. I. and Borowitzka, L. J. (1982) Symposium on taxonomic concepts in blue-green algae: towards a compromise with the bacterial code? *Taxon*, **31**, 673–683.

Frohlich, M. W. (1987) Common-is-primitive: A partial validation by tree counting. *Syst. Bot.*, **12**, 217–237.

Funk, V. A. and Brooks, D. R. (eds) (1981) *Advances in Cladistics*, New York Botanical Garden, New York.

Futuyma, D. J. (1986) *Evolutionary Biology (2nd edn)*, Sinauer Associates, Sunderland, Massachusetts.

Gaston, K. J. (1991) Body size and probability of description: the beetle fauna of Britain. *Ecol. Ent.*, **16**, 505–508.

Gaston, K. J. and May, R. M. (1992) Taxonomy of taxonomists. *Nature*, **356**, 281–282.

Gastony, G. J. (1991) Gene silencing in a polyploid homosporous fern: palaeopolyploidy revisited. *Proc. Natl. Acad. Sci. USA*, **88**, 1602–1605.

Gee, H. (1988) Friends and relations. *Nature*, **334**, 13–14.

Geist, V. (1992) Endangered species and the law. *Nature*, **357**, 274–276.

Georgiadis, N. J., Kat, P. W. and Oketch, H. (1990) Allozyme divergence within the Bovidae. *Evolution*, **44**, 2135–2149.

Gibbons, A. (1992) Mitochondrial Eve: wounded, but not dead yet. *Science*, **257**, 873–875.

Gillet, E. M. (1991) Genetic analysis of nuclear DNA restriction fragment patterns. *Genome*, **34**, 693–703.

Gledhill, D. (1985) *The Names of Plants*, Cambridge University Press, Cambridge.

Goldman, N. and Barton, N. H. (1992) Genetics and geography. *Nature*, **357**, 440–441.

Golenberg, E. M., Giannasi, D. E., Clegg, M. T., Smiley, C. J., Durbin, M., Henderson, D. and Zurawski, G. (1990) Chloroplast DNA sequence from Miocene magnolia species. *Nature*, **344**, 656–658.

Goodfellow, M., Jones, D. and Priest, F. G. (eds) (1985) *Computer-assisted Bacterial Systematics*, Academic Press, London.

Gosden, J. R., Mitchell, A. R., Seuanez, H. N. and Gosden, C. M. (1977) The distribution of sequences complementary to human satellite DNA's I, II and IV in the chromosomes of chimpanzee (*Pan troglodytes*), gorilla (*Gorilla gorilla*) and orangutan (*Pongo pygmaeus*). *Chromosome (Berl.)*, **63**, 253–271.

Gower, J. C. (1971) A general coefficient of similarity and some of its properties. *Biometrics*, **27**, 857–874.

Grantham, R., Perrin, P. and Mouchiroud, D. (1986) Patterns of codon usage of different kinds of species. In *Oxford Surveys in Evolutionary Biology. Volume 3*, R. Dawkins and M. Ridley (eds), Oxford University Press, Oxford, pp. 48–81.

Greenstone, M. H., Stuart, M. K. and Haunerland, N. H. (1991) Using monoclonal antibodies for phylogenetic analysis: an example from the Heliothinae (Lepidoptera: Noctuidae). *Ann. Entomol. Soc. Amer.*, **84**, 457–464.

Gutteridge, C. S., Vallis, L. and MacFie, H. J. H. (1985) Numerical methods in the classification of microorganisms by pyrolysis mass spectrometry. In *Computer-assisted Bacterial Systematics*, M. Goodfellow, D. Jones and F. G. Priest (eds), Academic Press, London, pp. 369–401.

Guyer, C. and Slowinski, J. B. (1991) Comparison of observed phylogenetic topologies with null expectations among three monophyletic lineages. *Evolution*, **45**, 340–350.

Gyllenberg, H. G. (1965) A model for computer identification of micro-organisms. *J. Gen. Microbiol.*, **39**, 401–405.

Haber, J. E. and Koshland, D. E. (1970) An evaluation of the relatedness of proteins based on comparison of amino acid sequences. *J. Mol. Biol.*, **50**, 617–639.

Hadrys, H., Balick, M. and Schierwater, B. (1992) Applications of random amplified polymorphic DNA (RAPD) in molecular ecology. *Molec. Ecol.*, **1**, 55–63.

Hall, A. V. (1970) A computer-based system for forming identification keys. *Taxon*, **19**, 12–18.

Hallam, A. (1988) The contribution of palaeontology to systematics and evolution. In *Prospects in Systematics. Systematics Association Special Volume No. 36*, D. L. Hawksworth (ed.), Clarendon Press, Oxford, pp. 128–147.

Harborne, J. B. (1968) The use of secondary chemicals in the systematics of higher plants. In *Chemotaxonomy and Serotaxonomy. Systematics Association Special Volume No. 2*, J. G. Hawkes (ed.), Academic Press, London, pp. 173–191.

Harding, E. F. (1971) The probabilities of rooted tree-shapes generated by random bifurcation. *Adv. Appl. Prob.*, **3**, 44–77.

Harris, H. and Hopkinson, D. A. (1976 *et seq*) *Handbook of Enzyme Electrophoresis in Human Genetics*, North Holland, Amsterdam.

Harvey, P. H. and Cotgreave, P. (1991) Avian phylogeny and distribution. (Book review). *Trends Ecol. Evol.*, **6**, 268–269.

Harvey, P. H. and Pagel, M. D. (1991) *The Comparative Method in Evolutionary Biology*, Oxford University Press, Oxford.

Haskell, P. T. and Morgan, P. J. (1988) User needs in systematics and obstacles to their fulfilment. In *Prospects in Systematics. Systematics Association Special Volume No. 36*, D. L. Hawksworth (ed.), Clarendon Press, Oxford, pp. 400–413.

Hawksworth, D. L. (ed.) (1988) *Prospects in Systematics. Systematics Association Special Volume No. 36*, Clarendon Press, Oxford.

Hawksworth, D. L. (1991) The fungal dimension of biodiversity: magnitude, significance, and conservation. Presidential address 1990. *Mycol. Res.*, **95**, 641–655.

Hawksworth, D. L. and Bisby, F. A. (1988) Systematics: the keystone of biology. In *Prospects in Systematics. Systematics Association Special Volume No. 36*, D. L. Hawksworth (ed.), Clarendon Press, Oxford, pp. 3–30.

Hawksworth, D. L., Sutton, B. C. and Ainsworth, G. C. (1983) *Ainsworth & Bisby's Dictionary of the Fungi (including the Lichens)* (7th edn), Commonwealth Mycological Institute, CAB, Kew.

Haymer, D. S., McInnis, D. O. and Arcangeli, L. (1992) Genetic variation between strains of the Mediterranean fruit fly, *Ceratitis capitata*, detected by DNA fingerprinting. *Genome*, **35**, 528–533.

Hendy, M. D. and Penny, D. (1982) Branch and bound algorithms to determine minimal evolutionary trees. *Math Biosci.*, **59**, 277–290.

Hendy, M. D., Steel, M. A., Penny, D. and Henderson, I. M. (1988) Families of trees and consensus. In *Classification and Related Methods of Data Analysis*, H. H. Bock (ed.), Elsevier Science Publishers, North Holland, pp. 355–362.

Hennig, W. (1950) *Grundzuge einer Theorie der phylogenetischen Systematik*, Deutscher Zentralverlag, Berlin.

Hennig, W. (1957) Systematik und Phylogenese. *Ber. Hundertj dtsch. ent. Ges.*, **1956**, 50–71.

Hennig, W. (1965) Phylogenetic systematics. *Ann. Rev. Entomol.*, **10**, 97–116.

Hennig, W. (1966) *Phylogenetic Systematics*, University of Illinois Press, Urbana.

Hennig, W. (1975) "Cladistic analysis or cladistic classification?": A reply to Ernst Mayr. *Syst. Zool.*, **24**, 244–256.

Heron, C. (1992) The networks of botanical creation. *New Scientist*, **1804**, 40–44.

Hey, J. (1992) Using phylogenetic trees to study speciation and extinction. *Evolution*, **46**, 627–640.

Hillis, D. M. (1987) Molecular versus morphological approaches to systematics. *Ann. Rev. Ecol. Syst.*, **18**, 23–42.

Hillis, D. M. and Dixon, M. T. (1991) Ribosomal DNA: molecular evolution and phylogenetic inference. *Q. Rev. Biol.*, **66**, 411–453.

Hillis, D. M. and Moritz, C. (eds) (1990) *Molecular Systematics*, Sinauer Associates, Sunderland, Massachusetts.

Hillis, D. M., Larson, A., Davis, S. K. and Zimmer, E. A. (1990) Nucleic acids III: sequencing. In *Molecular Systematics*, D. M. Hillis and C. Moritz (eds), Sinauer Associates, Sunderland, Massachusetts, pp. 318–370.

Hinegardner, R. and Rosen, D. E. (1972) Cellular DNA content and the evolution of teleostean fishes. *Amer. Nat.*, **106**, 621–644.

Horner, J. R., Varricchio, D. J. and Goodwin, M. B. (1992) Marine transgressions and the evolution of Cretaceous dinosaurs. *Nature*, **358**, 59–61.

Houde, P. (1987) Critical evaluation of DNA hybridization studies in avian systematics. *Auk*, **104**, 17–32.

Howard, J. C. (1988) How old is a polymorphism? *Nature*, **332**, 588–590.

Howard, R. W., Akre, R. D. and Garnett, W. B. (1990) Chemical mimicry in an obligate predator of carpenter ants (Hymenoptera: Formicidae). *Ann. Entomol. Soc. Amer.*, **83**, 607–617.

Hoyle, F. and Wickramasinghe, N. (1986) *Archaeopteryx, the Primordial Bird: A Case of Fossil Forgery*, Christopher Davies, Swansea.

Hughes, A. L. (1992) Avian species described on the basis of DNA only. *Trends Ecol. Evol.*, **7**, 2–3.

Hull, D. L. (1979) The limits of cladism. *Syst. Zool.*, **28**, 416–440.

Humphries, C. J. and Funk, V. A. (1984) Cladistic methodology. In *Current Concepts in Plant Taxonomy. Systematics Association Special Volume 25*, V. H. Heywood and D. M. Moore (eds), Academic Press, London, chapter 17, pp. 323–362.

Humphries, C. J. and Ladiges, P. Y. (1988) The application of cladistic taxonomy to plants: a response to Wilson. *Taxon*, **37**, 388–390.

Humphries, C. J. and Parenti, L. (1986) *Cladistic Biogeography*, Academic Press.

Huxley, J. (1958) Evolutionary processes and taxonomy with special reference to grades. *Uppsala Univ. Arssks.*, 21–38.

Imai, H. T. and Taylor, R. W. (1989) Chromosomal polymorphism involving telomere fusion, centromeric inactivation and centromere shift in the ant *Myrmecia* (*pilosula*) n=1. *Chromosoma*, **98**, 456–460.

Innis, M. A., Gelfand, D. H., Sninsky, J. J. and White, T. J. (eds) (1990) *PCR Protocols*, Academic Press, San Diego, London.

International Commission on Zoological Nomenclature (1985) *International Code of Zoological Nomenclature* (3rd edn), University of California Press, Berkeley and Los Angeles.

Iverson, T. H. and Flood, P. R. (1970) The morphology, occurrence and distributions of dilated cisternae of the endoplasmic reticulum in tissues of plants of the Cruciferae. *Protoplasma*, **71**, 467–477.

Jackson, R. C. (1971) The karyotype in systematics. *Ann. Rev. Ecol. Syst.*, **2**, 327–368.

Jackson, R. C. (1976) Evolution and systematic significance of polyploidy. *Ann. Rev. Ecol. Syst.*, **7**, 209–234.

Jackson, R. C. (1982) Polyploidy and diploidy: new perspectives on chromosome pairing and its evolutionary implication. *Amer. J. Bot.*, **69**, 1512–1523.

Jackson, R. C. (1984) Chromosome pairing in species and hybrids. In *Plant Biosystematics*, W. F. Grant (ed.), Academic Press, Toronto, pp. 67–86.

Jaeger, E. C. (1955) *A Source-Book of Biological Names and Terms* (3rd edn), Charles C Thomas, Springfield, Illinois.

Jamieson, B. G. M. (1987) *The Ultrastructure and Phylogeny of Insect Spermatozoa*, Cambridge University Press, Cambridge.

Jamieson, B. G. M. (1991) *Fish Evolution and Systematics: Evidence from Spermatozoa*, Cambridge University Press, Cambridge.

Jansen, R. K., Holsinger, K. E., Michaels, H. J. and Palmer, J. D. (1990) Phylogenetic analysis of chloroplast DNA restrictions site data at higher taxonomic levels: An example from the Asteraceae. *Evolution*, **44**, 2089–2105.

Jeffrey, C. (1982) *An Introduction to Plant Taxonomy* (2nd edn), Cambridge University Press, Cambridge.

Jeffrey, C. (1989) *Biological Nomenclature* (3rd edn), Edward Arnold, London.

Jeffreys, A. J., Wilson, V. and Thein, S. L. (1985) Hypervariable 'minisatellite' regions in human DNA. *Nature*, **314**, 67–72.

Jeffreys, A. J., Neumann, R. and Wilson, V. (1990) Repeat unit sequence variation in minisatellites: a novel source of DNA polymorphism for studying variation and mutation by molecular analysis. *Cell*, **60**, 473–485.

Jeffreys, A. J., MacLeod, A., Tamaki, K., Neil, D. L. and Monckton, D. G. (1991) Minisatellite repeat coding as a digital approach to DNA typing. *Nature*, **354**, 204–209.

Jelnes, J. E. (1986) Experimental taxonomy of *Bulinus* (Gastropoda: Planorbidae): the West and North African species reconsidered, based upon an electrophoretic study of several enzymes per individual. *Zool. J. Linn. Soc.*, **87**, 1–26.

Joysey, K. A. (1988) Some implications of the revolution in molecular biology. In *Prospects in Systematics. Systematics Association Special Volume No. 36*, D. L. Hawksworth (ed.), Clarendon Press, Oxford, pp. 202–216.

Jurzysta, M., Small, E. and Nozzolillo, C. (1988) Hemolysis, a synapomorphic discriminator of an expanded genus *Medicago* (Leguminosae). *Taxon*, **37**, 354–363.

Kazmer, D. J. (1991) Isoelectric focusing procedures for the analysis of allozymic variation in minute arthropods. *Ann. Entomol. Soc. Amer.*, **84**, 332–339.

Kemp, T. S. (1986) Feathered flights of fancy. *Nature*, **324**, 185.

Kemp, T. S. (1988) Haemothermia or Archaeosauria? The interrelationships of mammals, birds and crocodiles. *Zool. J. Linn. Soc.*, **92**, 67–104.

Kimura, M. (1983) The neutral theory of molecular evolution. In *Evolution of Genes and Proteins*, M. Nei and R. K. Koehn (eds), Sinauer, Sunderland, MA, pp. 208–233.

Kingsbury, D. W. (1986) Nomenclature of plant viruses. In *Biological Nomenclature Today*, W. D. L. Ride and T. Younès (eds), IRL Press, Eynsham, Oxford, pp. 49–53.

Kirby, L. T. (1990) *DNA Fingerprinting*, Stockton Press, New York.

Kluge, A. G. and Farris, J. S. (1969) Quantitative phyletics and the evolution of anurans. *Syst. Zool.*, **18**, 1–32.

Koop, B. F., Goodman, M., Xu, P., Chan, K. and Slightom, J. L. (1986) Primate Γ-globin DNA sequences and man's place among the great apes. *Nature*, **319**, 234–238.

Krieg, N. R. and Holt, J. G. (1984) *Bergey's Manual of Systematic Bacteriology. Volume 1*, Williams and Wilkins, Baltimore.

Krimbas, C. B. and Sourdis, J. (1987) Recent improvements in handling allelic isozyme data for tree construction. *Curr. Topics Biol. Med. Res. Vol. 15: Genetics development and evolution*, pp. 49–62.

Kwak, M. M. (1980) Artificial and natural hybridization and introgression in *Rhinanthus* (Scrophulariaceae) in relation to bumblebee pollination. *Taxon*, **29**, 613–628.

Lake, J. A. (1987) A rate-independent technique for analysis of nucleic acid sequences: evolutionary parsimony. *Mol. Biol. Evol.*, **4**, 167–191.

Lapage, S. P. (1971) Culture collections of bacteria. *Biol. J. Linn. Soc.*, **3**, 197–210.
Lapage, S. P., Sneath, P. H. A., Lessel, E. F., Skerman, V. B. D., Seeliger, H. P. R. and Clark, W. A. (eds) (1975) *International Code of Nomenclature of Bacteria (1976 Revision)*, American Society for Microbiology, Washington D.C.
Larson, A. (1989) The relationship between speciation and morphological evolution. In *Speciation and its Consequences*, D. Otte and J. A. Endler (eds), Sinauer Associates, Sunderland, Massachusetts, chapter 23, pp. 579–598.
Lavin, M., Doyle, J. J. and Palmer, J. D. (1990) Evolutionary significance of the loss of the chloroplast-DNA inverted repeat in the Leguminoseae subfamily Papilionoidea. *Evolution*, **44**, 390–402.
Legendre, P. (1975) *A posteriori* weighting of descriptors. *Taxon*, **24**, 603–608.
Leitch, A. R., Schwarzacher, T., Mosgöller, W., Bennett, M. D. and Heslop-Harrison, J. S. (1991) Parental genomes are separated throughout the cell cycle in a plant hybrid. *Chromosoma*, **101**, 206–213.
Le Quesne, W. J. (1964) A method of selection of characters in numerical taxonomy. *Syst. Zool.*, **18**, 201–205.
Levy, M. (1977) Minimum biosynthetic-step indices as measures of comparative flavonoid affinity. *Syst. Bot.*, **2**, 89–98.
Lidén, M. and Oxelman, B. (1989) Species – Pattern or process? *Taxon*, **38**, 228–232.
Lindahl, T. (1993) Instability and decay of the primary structure of DNA. *Nature*, **362**, 709–715.
Linde-Laursen, I. and von Bothmer, R. (1988) Elimination and duplication of particular *Hordeum vulgare* chromosomes in aneuploid interspecific *Hordeum* hybrids. *Theor. Appl. Gen.*, **76**, 897–908.
Löve, A. and Löve, D. (1961) Chromosome numbers of central and northwestern European plant species. *Opera Botanica Lund*, **5**, 1–581.
Ludwig, W., Wallner, G., Tesch, A. and Klink, F. (1991) A novel eubacterial phylum: comparative nucleotide sequence analysis of a *tuf*-gene of *Flexistipes sinusarabici*. *FEMS Microbiol. Letts*, **78**, 139–144.
Lundberg, J. G. (1972) Wagner networks and ancestors. *Syst. Zool.*, **21**, 398–413.
Lyster, S. (1985) *International Wildlife Law*, Grotius Press, Cambridge.
Macgregor, H. C. and Sessions, S. K. (1986) The biological significance of variation in satellite DNA and heterochromatin in newts of the genus *Triturus*: an evolutionary perspective. *Phil. Trans. R. Soc. Lond. B*, **312**, 243–259.
Macgregor, H. C. and Varley, J. M. (1983) *Working with Animal Chromosomes*, John Wiley & Sons, Chichester.
Maddison, W. (1989) Reconstructing character evolution on polytomous cladograms. *Cladistics*, **5**, 365–377.
Maddison, W. P. and Slatkin, M. (1991) Null models for the number of evolutionary steps in a character on a phylogenetic tree. *Evolution*, **45**, 1184–1197.
Maddison, W. P., Donoghue, M. J. and Maddison, D. R. (1984) Outgroup analysis and parsimony. *Syst. Zool.*, **33**, 83–103.
Margulis, L. (1988) Systematics: the view from the origin and early evolution of life. Secession of the Protista from animal and plant kingdoms. In *Prospects in Systematics. Systematics Association Special Volume No. 36*, D. L. Hawksworth (ed.), Clarendon Press, Oxford, pp. 430–443.
Margush, T. and McMorris, F. R. (1981) Consensus n-trees. *Bull. Math. Biol.*, **43**, 239–244.
Martin, A. P., Naylor, G. J. P. and Palumbi, S. R. (1992) Rates of mitochondrial DNA evolution in sharks are slow compared with mammals. *Nature*, **357**, 153–155.
Matthews, R. E. F. (1979) Classification and nomenclature of viruses: third report of the International Committee on Taxonomy of Viruses. *Intervirology*, **12**, 129–296.
Maxson, R. D. and Maxson, L. R. (1986) Microcomplement fixation: a quantitative estimator of protein evolution. *Mol. Biol. Evol.*, **3**, 375–388.
Maxson, L. R. and Maxson, R. D. (1990) Proteins II: immunological techniques. In

Molecular Systematics, D. M. Hillis and C. Moritz (eds), Sinauer Associates, Sunderland, Massachusetts.

May, R. M. (1988) How many species are there on earth? *Science*, **241**, 1441–1449.

May, R. M. (1992) Bottoms up for the oceans. *Nature*, **357**, 278–279.

Maynard Smith, J. (1989) *Evolutionary Genetics*, Oxford University Press, Oxford.

Mayr, E. (1942) *Systematics and the Origin of Species*, Columbia University Press, New York.

Mayr, E. (1965) Classification and phylogeny. *Amer. Zool.*, **5**, 165–174.

Mayr, E. (1969) *Principles of Systematic Zoology*, McGraw Hill, New York.

Mayr, E. (1990) A natural system of organisms. *Nature*, **348**, 491.

Mayr, E. (1992) A local flora and the biological species concept. *Amer. J. Bot.*, **79**, 222–238.

Mayr, E. and Ashlock, P. D. (1992) *Principles of Systematic Zoology* (2nd edn), McGraw Hill, New York.

Mayr, E. and Short, L. L. (1970) Species taxa of North American birds. *Nuttall Ornith. Club*, Publ. No. 9, Cambridge, Mass.

McCafferty, W. P. (1991) Toward a phylogenetic classification of the Ephemeroptera (Insecta): a commentary on systematics. *Ann. Entomol. Soc. Amer.*, **84**, 343–360.

McDade, L. (1990) Hybrids and phylogenetic systems, I. Patterns of character expression in hybrids and their implication for cladistic analysis. *Evolution*, **44**, 1685–1700.

McGinley, R. J. (1989) Entomological collection management – Are we really managing? *Insect Collection News*, **2**, 19–24.

McLellan, T. and Inouye, L. S. (1986) The sensitivity of isoelectric focusing and electrophoresis in the detection of sequence differences in proteins. *Biochem. Genet.*, **24**, 571–577.

McLennan, D. A. (1991) Integrating phylogeny and experimental ethology: from pattern to process. *Evolution*, **45**, 1773–1789.

McLennan, D. A., Brooks, D. R. and McPhail, J. D. (1988) The benefits of communication between comparative ethology and phylogenetic systematics: a case study using gasterosteid fishes. *Can. J. Zool.*, **66**, 2177–2190.

McPherson, M. J., Quirke, P. and Taylor, G. R. (eds) (1991) *PCR A Practical Approach*, Oxford University Press, Oxford.

Meacham, C. A. (1984a) The role of hypothesized direction of characters in the estimation of evolutionary history. *Taxon*, **33**, 26–38.

Meacham, C. A. (1984b) Evaluating characters by character compatibility analysis. In *Cladistics: Perspectives on the Reconstruction of Evolutionary History*, T. Duncan and T. F. Stuessy (eds), Columbia University Press, New York, pp. 152–165.

Meacham, C. A. and Estabrook, G. F. (1985) Compatibility methods in systematics. *Ann. Rev. Ecol. Syst.*, **16**, 431–446.

Michener, C. D. (1977) Discordant evolution and the classification of allodapine bees. *Syst. Zool.*, **26**, 32–56.

Mickevich, M. F. (1978) Taxonomic congruence. *Syst. Zool.*, **27**, 143–158.

Mickevich, M. F. (1982) Transformation series analysis. *Syst. Zool.*, **31**, 461–478.

Mishler, B. D. (1985) The morphological, developmental and phylogenetic basis for species concepts in bryophytes. *The Bryologist*, **88**, 207–214.

Mishler, B. D., Bremer, K., Humphries, C. J. and Churchill, S. P. (1988) The use of nucleic acid sequence data in phylogenetic reconstruction. *Taxon*, **37**, 391–395.

Mitter, C., Farrell, B. and Futuyma, D. J. (1991) Phylogenetic studies of insect-plant interactions: insights into the genesis of diversity. *Trends Ecol. Evol.*, **6**, 290–293.

Moritz, C. and Hillis, D. M. (1990) Molecular systematics: contents and controversies. In *Molecular Systematics*, D. M. Hillis and C. Moritz (eds), Sinauer Associates, Sunderland, Massachusetts.

Moritz, C., Dowling, T. E. and Brown, W. M. (1987) Evolution of animal mitochondrial DNA: relevance for population biology and systematics. *Ann. Rev. Ecol. Syst.*, **18**, 269–292.

Morse, L. E. (1974a) Computer programs for specimen identification, key construction and description printing. *Publ. Museum Michican State Univ. Biol.*, **5**.

Morse, L. E. (1974b) Computer-assisted storage and retrieval of the data of taxonomy and systematics. *Taxon*, **23**, 29–43.

Müller, G. B. and Wagner, G. P. (1991) Novelty in evolution: restructuring the concept. *Ann. Rev. Ecol. Syst.*, **22**, 229–256.

Mullis, K. B. and Faloona, F. A. (1987) Specific synthesis of DNA *in vitro* via a polymerase-catalysed chain reaction. *Methods Enzymol.*, **155**, 335–350.

Murphy, R. W., Sites, J. W. Jr., Buth, D. G. and Haufler, C. H. (1990) Proteins I: isozyme electrophoresis. In *Molecular Systematics*, D. M. Hillis and C. Moritz (eds), Sinauer Associates, Sunderland, Massachusetts.

Myers, A. A. and Giller, P. S. (1988) *Analytical Biogeography*, Chapman & Hall, London.

Navidi, W. C., Churchill, G. A. and von Haeseler, A. (1991) Methods for inferring phylogenies from nucleic acid sequence data by using maximum likelihood and linear invariants. *Mol. Biol. Evol.*, **8**, 128–143.

Nee, S. and Harvey, P. H. (1990) Profiting from biodiversity. *Nature*, **348**, 683.

Neff, N. A. (1986) A rational basis for *a priori* character weighting. *Syst. Zool.*, **35**, 110–123.

Nei, M. (1972) Genetic distance between populations. *Amer. Nat.*, **106**, 283–292.

Nei, M. (1978) Estimation of average heterozygosity and genetic distance from a small number of individuals. *Genetics*, **89**, 583–590.

Nelson, G. (1971) Paraphyly and polyphyly: redefinitions. *Syst. Zool.*, **20**, 471–472.

Nelson, G. (1974) Classification as an expression of phylogenetic relationship. *Syst. Zool.*, **22**, 344–359.

Nelson, G. (1978) Ontogeny, phylogeny, paleontology, and the biogenic law. *Syst. Zool.*, **27**, 324–345.

Nelson, G. (1979) Cladistic analysis and synthesis: principles and definitions, with a historical note on Adanson's *Familles des Plantes* (1763–1764). *Syst. Zool.*, **28**, 1–21.

Nelson, G. and Platnick, N. (1981) *Systematics and Biogeography: Cladistics and Vicariance*, Columbia University Press, New York.

Newell, I. M. (1970) Construction and use of tabular keys. *Pacific Ins*, **12**, 25–37.

Nicolson, D. H. (1980) Key to identification of effectively/ineffectively published material. *Taxon*, **29**, 485–488.

Nixon, K. C. and Davis, J. I. (1991) Polymorphic taxa, missing values and cladistic analysis. *Cladistics*, **7**, 233–241.

Nixon, K. C. and Wheeler, Q. D. (1992) Extinction and the origin of species. In *Extinction and Phylogeny*, M. J. Novacek and Q. D. Wheeler (eds), Columbia University Press, New York, pp. 119–143.

Nordal, I. (1987) Cladistics and character weighting: a contribution to the compatibility versus parsimony discussion. *Taxon*, **36**, 59–60.

Novacek, M. J. (1992) Fossils as critical data for phylogeny. In *Extinction and Phylogeny*, M. J. Novacek and Q. D. Wheeler (eds), Columbia University Press, New York, pp. 46–88.

Novacek, M. J. and Wheeler, Q. D. (eds) (1992) *Extinction and Phylogeny*, Columbia University Press, New York.

Noyes, J. S. (1989) A study of five methods of sampling Hymenoptera (Insecta) in a tropical rain forest with special reference to Parasitica. *J. Nat. Hist.*, **23**, 285–298.

Nuttall, G. H. F. (1904) *Blood Immunity and Blood Relationships. A demonstration of certain blood relationships among animals by means of the precipitin test for blood*, Cambridge University Press, Cambridge, UK.

Ogihara, Y. and Tsunewaki, K. (1988) Diversity and evolution of chloroplast DNA in *Triticum* and *Aegilops* as revealed by restriction fragment analyses. *Theor. Appl. Genet.*, **76**, 321–332.

Orr, H. A. (1990) "Why polyploidy is rarer in animals than in plants" revisited. *Amer. Nat.*, **136**, 759–770.

Osborne, D. V. (1963) Some aspects of the theory of dichotomous keys. *New Phytol.*, **62**, 144–160.

O'Toole, C. and Raw, A. (1991) *Bees of the World*, Blandford, London.

Ouchterlony, O. (1953) Antigen-antibody reaction in gels: types of reactions in co-ordinated systems of diffusion. *Acta Path. Microbiol. Scand.*, **32**, 231–240.

Ouchterlony, O. (1958) Diffusion-in-gel methods for immunological analysis. *Prog. Allergy*, **5**, 1.

Palmer, J. D. (1985) Comparative organization of chloroplast genomes. *Annu. Rev. Genet.*, **19**, 325–354.

Pamilo, P., Pekkarinen, A. and Varvio-aho, S.-L. (1981) Phylogenetic relationships and the origin of social parasitism in Vespidae and in *Bombus* and *Psithyrus* as revealed by enzyme genes. In *Biosystematics of Social Insects. Systematics Association Special Volume 19*, P. E. Howse and J.-L. Clément (eds), Academic Press, London, pp. 37–61.

Panchen, A. L. (1992) *Classification, Evolution, and the Nature of Biology*, Cambridge University Press, Cambridge.

Pankhurst, R. J. (1970) A computer program for generating diagnostic keys. *Comp. J.*, **13**, 145–151.

Pankhurst, R. J. (1971) Botanical keys generated by computer. *Watsonia*, **8**, 357–368.

Pankhurst, R. J. (1974) Automated identification in systematics. *Taxon*, **23**, 45–51.

Pankhurst, R. J. (ed.) (1975) *Biological Identification with Computers. Systematics Association Special Volume 7*, Academic Press, London.

Pankhurst, R. J. (1978) *Biological Identification*, Edward Arnold, London.

Pankhurst, R. J. (1983) The construction of a floristic database. *Taxon*, **32**, 193–202.

Pankhurst, R. J. (1988a) An interactive program for the construction of identification keys. *Taxon*, **37**, 747–755.

Pankhurst, R. J. (1988b) Database design for monographs and floras. *Taxon*, **37**, 733–746.

Pankhurst, R. J. (1991) *Practical Taxonomic Computing*, Cambridge University Press, Cambridge, UK.

Pankhurst, R. J. and Aitchison, R. R. (1975) A computer program to construct polyclaves. In *Biological Identification with Computers. Systematics Association Special Volume 7*, R. J. Pankhurst (ed.), Academic Press, London, pp. 73–78.

Partridge, T. R., Dallwitz, M. J. and Watson, L. (1988) *A Primer for the DELTA System on MS-DOS and VMS (2nd edn), CSIRO, Division of Entomology Report No. 38*, CSIRO, Canberra, Australia.

Pashley, D. P., Hammond, A. M. and Hardy, T. N. (1992) Reproductive isolating mechanisms in Fall Armyworm host strains (Lepidoptera: Noctuidae). *Ann. Entomol. Soc. Amer.*, **85**, 400–405.

Patterson, C. (1981) The significance of fossils in determining evolutionary relationships. *Ann. Rev. Ecol. Syst.*, **12**, 195–223.

Patterson, C. (1982) Morphological characters and homology. In *Problems of Phylogenetic Reconstruction*, K. A. Joysey and A. E. Friday (eds), Academic Press, London, pp. 21–74.

Patterson, C. (1990) Reassessing relationships. *Nature*, **344**, 199–200.

Patterson, C. and Rosen, D. E. (1977) Review of ichthyodectiform and other mesozoic teleost fishes and the theory and practice of classifying fossils. *Bull. Am. Mus. Nat. Hist.*, **158**, 81–172.

Payne, R. W. (1975) Genkey: A program for constructing diagnostic keys by R. W. Payne. In *Biological Identification with Computers. Systematics Association Special Volume 7*, R. J. Pankhurst (ed.), Academic Press, London, pp. 65–72.

Penny, D. and Hendy, M. D. (1985) Testing methods of evolutionary tree construction. *Cladistics*, **1**, 266–278.

Penny, D., Foulds, L. R. and Hendy, M. D. (1982) Testing the theory of evolution by comparing phylogenetic trees constructed from five different protein sequences. *Nature*, **297**, 197–200.

Penny, D., Hendy, M. D. and Steel, M. A. (1992) Progress with methods for constructing evolutionary trees. *Trends Ecol. Evol.*, **7**, 73–79.

Pesole, G., Bozzetti, M. P., Lanave, C., Preparata, G. and Saccone, C. (1991) Glutamine synthetase gene evolution: a good molecular clock. *Proc. Natl. Acad. Sci. USA*, **88**, 522–526.

Peterson, A. T. and Lanyon, S. M. (1992) New bird species, DNA studies and type specimens. *Trends Ecol. Evol.*, **7**, 167–168.

Phillips, R. B. (1983) Shape characters in numerical taxonomy and problems with ratios. *Taxon*, **32**, 535–544.

Platnick, N. I. (1976) Are monotypic genera possible? *Syst. Zool.*, **25**, 198–199.

Platnick, N. I. (1977a) Paraphyletic and polyphyletic groups. *Syst. Zool.*, **26**, 195–200.

Platnick, N. I. (1977b) Cladograms, phylogenetic trees, and hypothesis testing. *Syst. Zool.*, **26**, 438–442.

Platnick, N. I. (1979) Philosophy and the transformation of cladistics. *Syst. Zool.*, **28**, 537–546.

Platnick, N. I. (1987) An empirical comparison of microcomputer parsimony programs. *Cladistics*, **3**, 121–144.

Platnick, N. I. (1989) An empirical comparison of microcomputer parsimony programs, II. *Cladistics*, **5**, 145–161.

Pogue, M. J. and Mickevich, M. F. (1990) Character definitions and character state delineation: the *bête noire* of phylogenetic inference. *Cladistics*, **6**, 319–361.

Poinar, G. O. (1992) *Life in Amber*, Stanford University Press, Stanford.

Powell, A. M. (1985) Crossing data as generic criteria in the Asteraceae. *Taxon*, **34**, 55–60.

Quicke, D. L. J. (1988) Spiders bite their way towards safer insecticides. *New Scientist*, **1640**, 38–41.

Quicke, D. L. J. and Usherwood, P. N. R. (1990) Spider toxins as lead structures for novel pesticides. In *Safer Insecticides: Development and Use*, E. Hodgson and R. J. Kuhr (eds), Marcel Dekker Inc., New York, pp. 385–452.

Rasnitsyn, A. (1982) Proposal to regulate the names of taxa above the family group. Z.N.(S.) 2381. *Bull. Zool. Nom.*, **39**, 200–207.

Remane, A. (1956) Die Grundlagen des natürlichen Systems, vergleichenden Anatomie und der Phylogenetik. *Theoretische Morphologie und Systematik, Volume 1 (2nd edn)*, Akademische Verlagsgesellschaft Geest & Portig, Leipzig.

Ricciardi, A. and Reiswig, H. M. (1992) *Spongilla heterosclerifera* Smith, 1918, is an interspecific freshwater sponge mixture (Porifera, Spongillidae). *Can. J. Zool.*, **70**, 352–354.

Richardson, B. J., Baverstock, P. R. and Adams, M. (1986) *Allozyme Electrophoresis. A Handbook for Animal Systematics and Population Studies*, Academic Press, San Diego.

Ride, W. D. L. (1988) Towards a unified system of biological nomenclature. In *Prospects in Systematics. Systematics Association Special Volume No. 36*, D. L. Hawksworth (ed.), Clarendon Press, Oxford, pp. 332–353.

Riding, R. (1992) The algal breath of life. *Nature*, **359**, 13–14.

Rieger, R. and Tyler, S. (1979) The homology theorem in ultrastructural research. *Amer. Zool.*, **19**, 655–664.

Robinson, H. (1986) A key to the common errors of cladistics. *Taxon*, **35**, 309–311.

Rogers, J. S. (1972) Measures of genetic similarity and genetic distance. *Univ. Texas Publs*, **7213**, 145–153.

Rohlf, F. J. (1963) Congruence of larval and adult classifications in *Aedes* (Diptera: Culicidae). *Syst. Zool.*, **12**, 97–117.

Rohlf, F. J. (1974) Methods of comparing classifications. *Ann. Rev. Ecol. Syst.*, **5**, 101–113.

Rohlf, F. J. (1982) Consensus indices for comparing classifications. *Math. Biosci.*, **59**, 131–144.

Rohlf, F. J. and Wooten, M. C. (1988) Evaluation of the restricted maximum-likelihood method for estimating phylogenetic allele-frequency data. *Evolution*, **42**, 581–595.

Rohlf, F. J., Chang, W. S., Sokal, R. R. and Kim, J. (1990) Accuracy of estimating phylogenies: effects of tree topology and evolutionary model. *Evolution*, **44**, 1671–1684.

Rohrer, J. R. (1988) Incongruence between gametophytic and sporophytic classifications in mosses. *Taxon*, **37**, 838–845.

Rosen, D. E. (1978) Vicariant patterns and historical explanations in biogeography. *Syst. Zool.*, **27**, 159–188.

Rosen, D. (1979) Fishes from the upland intermontane basins of Guatemala. *Bull. Amer. Mus. Nat. Hist.*, **162**, 269–375.

Rosengren, R. (1981) The pupa-carrying test as a taxonomic tool in the *Formica rufa* group. In *Biosystematics of Social Insects. Systematics Association Special Volume 19*, P. E. Howse and J.-L. Clément (eds), Academic Press, London, pp. 263–281.

Ross, K. G. and Carpenter, J. M. (1991) Phylogenetic analysis and the evolution of queen number in eusocial Hymenoptera. *J. Evol. Biol.*, **4**, 117–130.

Ryder, O. A., Kumamoto, A. T., Durrant, B. S. and Benirschke, K. (1989) Chromosomal divergence and reproductive isolation in dik-diks. In *Speciation and its Consequences*, D. Otte and J. A. Endler (eds), Sinauer Associates, Sunderland, Massachusetts, chapter 9, pp. 208–225.

Rypka, E. W. and Babb, R. (1970) Automatic construction and use of an identification scheme. *Med. Res. Eng.*, **9**, 9–19.

Sackin, M. J. (1985) Comparisons of classifications. In *Computer-assisted Bacterial Systematics*, M. Goodfellow, D. Jones and F. G. Priest (eds), Academic Press, London, pp. 21–36.

Saito, N. and Nei, M. (1987) The neighbor-joining method: a new method for reconstructing phylogenetic trees. *Mol. Biol. Evol.*, **4**, 406–425.

Salomon, M. (1989) Song as a possible reproductive isolating mechanism between two parapatric forms. The case of the chaffinch *Phylloscopus c. collybita* and *P. c. brehmii* in the Western Pyrenees. *Behaviour*, **111**, 270–290.

Sanderson, M. J. (1991) In search of homoplastic tendencies: statistical inference of topological pattern in homoplasy. *Evolution*, **45**, 351–358.

Sanderson, M. J. and Donoghue, M. J. (1989) Patterns of variation in levels of homoplasy. *Evolution*, **43**, 1781–1795.

Sargent, J. R. and George, S. G. (1975) *Methods of Zone Electrophoresis*, BDH Chemicals Ltd, Poole.

Sarich, V. M., Schmid, C. W. and Marks, J. (1989) DNA hybridization as a guide to phylogenies: a critical analysis. *Cladistics*, **5**, 3–32.

Scadding, S. R. (1981) Do 'vestigial organs' provide evidence for evolution? *Evol. Theory*, **5**, 173–176.

Schaeffer, B. (1987) Deuterostome monophyly and phylogeny. In *Evolutionary Biology*, Volume 21, M. K. Hecht, B. Wallace and G. T. Prance (eds), Plenum Press, New York, chapter 7, pp. 179–235.

Schleifer, K. H. and Stackebrandt, E. (1983) Molecular systematics of prokaryotes. *Ann. Rev. Microbiol.*, **37**, 143–187.

Schwarzacher, T., Leitch, A. R., Bennett, M. D. and Heslop-Harrison, J. S. (1989) *In situ* localization of parental genomes in a wide hybrid. *Ann. Bot.*, **64**, 315–324.

Scott, P. and Rines, R. (1975) Naming the Loch Ness Monster. *Nature*, **258**, 466.

Seabright, M. (1971) A rapid banding technique for human chromosomes. *Lancet*, **2**, 971–972.

Seaman, F. C. and Funk, V. A. (1983) Cladistic analysis of complex natural products: Developing transformation series for sesquiterpene lactone data. *Taxon*, **32**, 1–27.

Sederoff, R. R. (1984) Structural variation in mitchondrial DNA. *Adv. Genet.*, **22**, 1–108.

Selden, P. A. and Gall, J.-C. (1992) A triassic mygalomorph spider from the Northern Vosges, France. *Palaeontology*, **35**, 211–235.

Selden, P. A., Shear, W. A. and Bonam, P. M. (1991) A spider and other arachnids from the Devonian of New York, and a reinterpretation of Devonian Araneae. *Palaeontology*, **34**, 241–281.

Sharkey, M. J. (1989) A hypothesis-independent method of character weighting for cladistic analysis. *Cladistics*, **5**, 63–86.

Shaw, C. R. and Prasad, R. (1970) Starch gel electrophoresis of enzymes – a compilation of recipes. *Biochem. Genet.*, **4**, 297–330.

Sheldon, P. R. (1987) Parallel gradualistic evolution of Ordovician trilobites. *Nature*, **330**, 561–563.

Shields, G. F. and Helm-Bychowski, K. M. (1988) Mitochondrial DNA of birds. In *Current Ornithology, Volume 5*, R. F. Johnston (ed.), Plenum Press, New York, pp. 273–295.

Sibley, C. G. and Ahlquist, J. E. (1983) Phylogeny and classification of birds based on the data of DNA–DNA hybridization. In *Current Ornithology, Volume 1*, R. F. Johnston (ed.), Plenum Press, New York, pp. 245–292.

Sibley, C. G. and Ahlquist, J. E. (1986) Reconstructing bird phylogeny by comparing DNAs. *Scientific Amer.*, **254**, 68–78.

Sibley, C. G. and Ahlquist, J. E. (1990) *Phylogeny and Classification of Birds: A Study in Molecular Evolution*, Yale University Press.

Sigler, L. and Hawksworth, D. L. (1987) International Commission on the Taxonomy of Fungi. Code of practice for systematic mycologists. *Microbiol. Sci.*, **4**, 83–86; *Mycologist*, **21**, 101–105; *Mycopathologia*, **99**, 3–7.

Simpson, G. G. (1961) *Principles of Animal Taxonomy*, Columbia University Press, New York.

Sivarajan, V. V. (1991) *Introduction to the Principles of Plant Taxonomy* (2nd edn), N. K. B. Robson (ed.), Cambridge University Press, Cambridge.

Skerman, V. B. D., McGowan, V. and Sneath, P. H. A. (1980) Approved lists of bacterial names. *Int. J. Syst. Bacteriol.*, **30**, 225–420.

Smith, A. B. (1992) Echinoderm phylogeny: morphology and molecules approach accord. *Trends Ecol. Evol.*, **7**, 224–229.

Smith, E. F. G., Arctander, P., Fjeldså, J. and Amir, O. G. (1991) A new species of shrike (Laniidae: *Laniarius*) from Somalia, verified by DNA sequence data from the only known individual. *Ibis*, **133**, 227–235.

Smith, M. W., Feng, D.-F. and Doolittle, R. F. (1992) Evolution by acquisition: the case for horizontal gene transfers. *Trends Biochem. Sci.*, **17**, 489–493.

Smith, P. G. and Phipps, J. B. (1984) Consensus trees in phenetic analyses. *Taxon*, **33**, 586–594.

Smith, P. M. (1976) *The Chemotaxonomy of Plants*, Edward Arnold, London.

Sneath, P. H. A. (1985) Future of numerical taxonomy. In *Computer-assisted Bacterial Systematics*, M. Goodfellow, D. Jones and F. G. Priest (eds), Academic Press, London, pp. 415–431.

Sneath, P. H. A. and Sokal, R. R. (1973) *Numerical Taxonomy*, W. H. Freeman, San Francisco, California.

Sober, E. (1988) *Reconstructing the Past. Parsimony, Evolution, and Inference*, The MIT Press, Cambridge, Massachusetts.

Sokal, R. R. and Crovello, T. J. (1970) The biological species concept: a critical evaluation. *Amer. Nat.*, **104**, 127–153.

Sokal, R. R. and Sneath, P. H. A. (1963) *Principles of Numerical Taxonomy*, W. H. Freeman & Co, San Francisco.

Song, K. and Osborn, T. C. (1992) Polyphyletic origins of *Brassica napus*: new evidence based on organelle and nuclear RFLP analyses. *Genome*, **35**, 992–1001.

Sorensen, J. T. (1990) Taxonomic partitioning of discrete-state phylogenies: Relationships of the aphid subtribes Eulachnina and Schizolachnina (Homoptera: Aphidiidae: Lachninae). *Ann. Entomol. Soc. Amer.*, **83**, 394–408.

Soulé, M. E. (1990) The real work of systematics. *Annals Missouri Bot. Gard.*, **77**, 4–12.

Stace, C. A. (1980) *Plant Taxonomy and Biosystematics*, Edward Arnold, London.

Stearn, W. T. (1973) *Botanical Latin*. David & Charles, Newton Abbott.

Steele, K. P., Holsinger, K. E., Jansen, R. K. and Taylor, D. W. (1988) Phylogenetic relationships in green plants – A comment on the use of 5S ribosomal RNA sequences by Bremer *et al. Taxon*, **37**, 135–138.

Steinemann, M., Pinsker, W. and Sperlich, D. (1984) Chromosome homologies within the *Drosophila obscura* group probed by *in situ* hybridization. *Chromosoma*, **91**, 46–53.

Stevens, P. F. (1980) Evolutionary polarity of character states. *Ann. Rev. Ecol. Syst.*, **11**, 333–358.

Storch, V. (1979) Contributions of comparative ultrastructural research to problems of invertebrate evolution. *Amer. Zool.*, **19**, 637–645.

Stouthamer, R., Lack, R. F. and Hamilton, W. D. (1990) Antibiotics cause parthenogenetic *Trichogramma* (Hymenoptera/Trichogrammatidae) to revert to sex. *Proc. U. S. Acad. Sci.*, **87**, 2424–2427.

Stuessy, F. and Crisci, J. V. (1984) Problems in the determination of evolutionary directionality of character-state change for phylogenetic reconstruction. In *Cladistics: Perspectives on the Reconstruction of Evolutionary History*, T. Duncan and T. F. Stuessy (eds), Columbia University Press, New York, pp. 71–87.

Sundberg, P. (1985) Nemertean systematics and phenetic classification: an example from a group of hoplonemerteans. *Zool. J. Linn. Soc.*, **85**, 247–266.

Sundberg, P. (1989) Phylogenetic and cladistic classification of terrestrial nemerteans: the genera *Pantinonemertes* Moore and Gibson and *Geonemertes* Semper. *Zool. J. Linn. Soc.*, **95**, 363–372.

Sutton, S. L. and Collins, N. M. (1991) Insects and tropical forest conservation. In *The Conservation of Insects and their Habitats*, N. M. Collins and J. A. Thomas (eds), Academic Press, London, pp. 405–424.

Swofford, D. L. (1981) On the utility of the distance Wagner procedure. In *Advances in Cladistics. Proceedings of the First Meeting of the Willi Hennig Society*, V. A. Funk and D. R. Brooks (eds), New York Botanical Gardens, Bronx, pp. 25–43.

Swofford, D. L. (1991) *PAUP: Phylogenetic Analysis Using Parsimony, Version 3.0*, Computer program distributed by the Illinois Natural History Survey, Champaign, Illinois.

Swofford, D. L. and Berlocher, S. H. (1987) Inferring evolutionary trees from gene frequency data under the principle of maximum parsimony. *Syst. Zool.*, **36**, 293–325.

Swofford, D. L. and Maddison, W. P. (1987) Reconstructing ancestral character states under Wagner parsimony. *Math. Biosci.*, **87**, 199–229.

Swofford, D. L. and Olsen, G. J. (1990) Phylogenetic reconstruction. In *Molecular Sytstematics*, D. M. Hillis and C. Moritz (eds), Sinauer Associates, Sunderland, Massachusetts.

Sytsma, K. J. (1990) DNA and morphology: inference of plant phylogeny. *Trends Ecol. Evol.*, **5**, 104–110.

Tehler, A. (1982) The species pair concept in lichenology. *Taxon*, **31**, 708–717.

Templeton, A. R. (1983) Phylogenetic inference from restriction endonuclease cleavage site maps with particular reference to the evolution of humans and apes. *Evolution*, **37**, 221–244.

Thompson, J. D. and Lumaret, R. (1992) The evolutionary dynamics of polyploid plants: Origins, establishment and persistence. *Trends Ecol. Evol.*, **7**, 302–307.

Thorpe, J. P. (1979) Enzyme variation and taxonomy: the estimation of sampling errors in measurements of interspecific genetic similarity. *J. Linn. Soc. (Biol.)*, **11**, 369–386.

Thorpe, J. P. (1982) The molecular clock hypothesis: biochemical evolution, genetic differentiation and systematics. *Ann. Rev. Ecol. Syst.*, **13**, 139–168.

Thorpe, J. P. (1983) Enzyme variation, genetic distance and evolutionary divergence in relation to taxonomic separation. In *Protein Polymorphism: Adaptive and Taxonomic Significance*, G. S. Oxford and D. Rollinson (eds), Academic Press, London, pp. 131–152.

Thorpe, R. S. (1984) Coding morphometric characters for constructing distance Wagner networks. *Evolution*, **38**, 244–255.

Tilling, S. M. (1984) Keys to biological identification: their role and construction. *J. Biol. Education*, **18**, 293–304.

Tilling, S. M. (1987) Education and taxonomy: the role of the Field Studies Council and AIDGAP. *Biol. J. Linn. Soc.*, **32**, 87–96.

Tyler, S. (1979) Contributions of electron microscopy to systematics and phylogeny: Introduction to the symposium. *Amer. Nat.*, **19**, 541–543.

Vane-Wright, R. I., Humphries, C. J. and Williams, P. H. (1991) What to protect – systematics and the agony of choice. *Biol. Conserv.*, **55**, 235–254.

Vavilov, N. I. (1922) The law of homologous series of variations. *J. Genet.*, **12**, 47–89.

Vorster, P. (1986) The use of the term paralectotype/lectoparatype. *Taxon*, **35**, 316–317.

Wagner, G. P. (1989) The biological homology concept. *Ann. Rev. Ecol. Syst.*, **20**, 51–69.

Wagner, W. H. Jr (1961) Problems in the classification of ferns. *Recent Adv. Bot.*, **1**, 841–844.

Wagner, W. H. (1980) Origin and philosophy of the groundplan divergence method of cladistics. *Syst. Bot.*, **5**, 173–193.

Wake, D. B. (1991) Homoplasy: the result of natural selection, or evidence for design limitations. *Amer. Nat.*, **138**, 543–567.

Wake, D. B., Yanev, K. P. and Frelow, M. M. (1989) Sympatry and hybridization in a "ring species": the plethodontid salamander *Ensatina eschscholtzii*. In *Speciation and its Consequences*, D. Otte and J. A. Endler (eds), Sinauer Associates, Sunderland, Massachusetts, chapter 6, pp. 134–157.

Walker, J. W. (1979) Contributions of electron microscopy to angiosperm phylogeny and systematics: significance of ultrastructural characters in delimiting higher taxa. Sieve plates and their plastid inclusions. *Amer. Zool.*, **19**, 609–619.

Walker, R. J. and Holden-Dye, L. (1989) Commentary on the evolution of transmitters, receptors and ion channels in invertebrates. *Comp. Biochem. Physiol. (A)*, **93**, 25–39.

Walters, S. M. (1975) Traditional methods of biological identification. In *Biological Identification with Computers. Systematics Association, Special Volume 7*, R. J. Pankhurst (ed.), Academic Press, London, pp. 3–8.

Walther, J. R. (1981) Cuticular sense organs as characters in phylogenetic research. *Mitt. dtsch. Ges. allg. angew. Ent.*, **3**, 146–150.

Wanntorp, H.-E., Brooks, D. R., Nilsson, T., Nylin, S., Ronqvist, F., Stearns, S. C. and Weddell, N. (1990) Phylogenetic approaches to ecology. *Oikos*, **57**, 119–132.

Watrous, L. E. and Wheeler, Q. D. (1981) The outgroup comparison method of character analysis. *Syst. Zool.*, **30**, 1–11.

Watson, L. and Milne, P. (1972) A flexible system for automatic generation of special-purpose dichotomous keys, and its application to Australian grass genera. *Aust. J. Bot.*, **20**, 331–352.

Welsh, J. and McClelland, M. (1990) Fingerprinting genomes using PCR with arbitrary primers. *Nucleic Acids Res.*, **18**, 7213–7218.

Wenzel, J. W. (1991) Evolution of nest architecture. In *The Social Biology of Wasps*, K. G. Ross and R. W. Matthews (eds), Comstock Publishing Associates, Ithaca, pp. 480–519.

Werman, S. D., Springer, M. S. and Britten, R. J. (1990) Nucleic acids 1: DNA–DNA hybridization. In *Molecular Systematics*, D. M. Hollis and C. Moritz (eds), Sinauer Associates, Sunderland, Massachusetts.

Westbroek, P., van der Meide, P. H., van der Wey-Kloppers, J. S., van der Sluis, R. J., de Leeuw, J. W. and de Jong, E. W. (1979) Fossil macromolecules from cephalopod shells: characterization, immunological response and diagenesis. *Palaeobiology*, **5**, 151–167.

Wheeler, Q. D. (1990a) Insect diversity and cladistic constraints. *Ann. Entomol. Soc. Amer.*, **83**, 1031–1047.

Wheeler, Q. D. (1990b) Ontogeny and character phylogeny. *Cladistics*, **6**, 225–268.

White, M. J. D. (1982) Rectangularity, speciation, and chromosome architecture. In *Mechanisms of Speciation*, C. Barigozzi (ed.), Alan R. Liss, New York, pp. 75–103.

Wildly, P. (1971) Classification and nomenclature of viruses: First report of the International Committee on Taxonomy of Viruses. *Monograph on Virology, Volume 5*, Karger, Basel.

Wiley, E. O. (1979) Cladograms and phylogenetic trees. *Syst. Zool.*, **28**, 88–92.

Wiley, E. O. (1981) *Phylogenetics: The Theory and Practice of Phylogenetic Systematics*, John Wiley & Sons, New York.

Williams, J. (1985) Cladistics and the evolution of proteins. In *Computer-assisted Bacterial Systematics*, M. Goodfellow, D. Jones and F. G. Priest (eds), Academic Press, London, pp. 61–89.

Williams, J. G. K., Kubelik, A. R., Livak, K. J., Rafalski, J. A. and Tingey, S. V. (1990) DNA polymorphisms amplified by arbitrary primers are useful genetic markers. *Nucleic Acids Res.*, **18**, 6531–6535.

Wilson, E. O. (1988) The current state of biological diversity. In *Biodiversity*, E. O. Wilson (ed.), National Academy Press, Washington D.C., pp. 3–18.

Wilson, J. B. (1988) The application of cladistic taxonomy to plants. *Taxon*, **37**, 387–388.

Wilson, J. B. and Partridge, T. R. (1986) Interactive plant identification. *Taxon*, **35**, 1–12.

d'Winter, A. J. and Rollenhagen, T. (1990) The importance of male and female acoustic behaviour for reproductive isolation in *Ribautodelphax* planthoppers (Homoptera: Delphacidae). *Biol. J. Linn. Soc.*, **40**, 191–206.

Yoshimoto, C. M. (1978) Voucher specimens for entomology in North America. *Ent. Soc. Amer. Bull.*, **24**, 141–142.

Young, J. P. W. (1988) The estimation of protein and nucleic acid homologies. In *Prospects in Systematics. Systematics Association Special Volume No. 36*, D. L. Hawksworth (ed.), Clarendon Press, Oxford, pp. 169–183.

Zarco, C. R. (1986) A new method for estimating karyotype asymmetry. *Taxon*, **35**, 526–530.

Zouros, E., Freeman, K. R., Oberhauser Ball, A. and Pogson, G. H. (1992) Direct evidence for extensive paternal inheritance in the marine mussel *Mytilus*. *Nature*, **359**, 412–414.

Index